Do Numbers Exist?

In *Do Numbers Exist?* Peter van Inwagen and William Lane Craig take opposite sides on whether there are abstract objects, such as numbers and properties. Craig argues that there are no abstract objects, whereas van Inwagen argues that there are. Their exchange explores various arguments about the existence and nature of abstract objects. They focus especially on whether our ordinary and scientific thought and talk commit us to abstract objects, surveying the options available to us and the objections each faces. The debate covers central problems and methods in metaphysics, and also delves into theological questions raised by abstract objects.

Key Features

- Showcases the presentation and defense of two points of view on the existence of abstract objects, from two of the world's leading philosophers
- Presents definitions in an easily accessible form
- Provides frequent summaries of previously covered material
- Includes a glossary of all specialized vocabulary

Peter van Inwagen is the John Cardinal O'Hara Professor of Philosophy Emeritus at the University of Notre Dame. He is the author of two seminal works, *An Essay on Free Will* (1983), and *Material Beings* (1990), both of which are still in print.

William Lane Craig is Professor of Philosophy at Talbot School of Theology and at Houston Christian University. He has authored or edited over 40 books, including *The* Kalām *Cosmological Argument*; *God, Time, and Eternity*; and *God and Abstract Objects*, as well as over 200 articles in professional publications of philosophy and theology.

Little Debates about Big Questions

About the series:

Philosophy asks questions about the fundamental nature of reality, our place in the world, and what we should do. Some of these questions are perennial: for example, *Do we have free will? What is morality?* Some are much newer: for example, *How far should free speech on campus extend? Are race, sex and gender social constructs?* But all of these are among the big questions in philosophy and they remain controversial.

Each book in the *Little Debates about Big Questions* series features two professors on opposite sides of a big question. Each author presents their own side, and the authors then exchange objections and replies. Short, lively, and accessible, these debates showcase diverse and deep answers. Pedagogical features include standard form arguments, section summaries, bolded key terms and principles, glossaries, and annotated reading lists.

The debate format is an ideal way to learn about controversial topics. Whereas the usual essay or book risks overlooking objections against its own proposition or misrepresenting the opposite side, in a debate each side can make their case at equal length, and then present objections the other side must consider. Debates have a more conversational and fun style too, and we selected particularly talented philosophers—in substance and style—for these kinds of encounters.

Debates can be combative—sometimes even descending into anger and animosity. But debates can also be cooperative. While our authors disagree strongly, they work together to help each other and the reader get clearer on the ideas, arguments, and objections. This is intellectual progress, and a much-needed model for civil and constructive disagreement.

The substance and style of the debates will captivate interested readers new to the questions. But there's enough to interest experts too. The debates will be especially useful for courses in philosophy and related subjects—whether as primary or secondary readings—and a few debates can be combined to make up the reading for an entire course.

We thank the authors for their help in constructing this series. We are honored to showcase their work. They are all preeminent scholars or rising-stars in their fields, and through these debates they share what's been discovered with a wider audience. This is a paradigm for public philosophy, and will impress upon students, scholars, and other interested readers the enduring importance of debating the big questions.

Tyron Goldschmidt, Fellow of the Rutgers Center for Philosophy of Religion, USA

Dustin Crummett, Ludwig Maximilian University of Munich, Germany

Published Titles:

Do We Have Free Will?: A Debate
By Robert Kane and Carolina Sartorio

Is There a God?: A Debate
By Kenneth L. Pearce and Graham Oppy

Is Political Authority an Illusion?: A Debate
By Michael Huemer and Daniel Layman

Selected Forthcoming Titles:

Should We Want to Live Forever?: A Debate
By Stephen Cave and John Martin Fischer

What Do We Owe Other Animals?: A Debate
By Bob Fischer and Anja Jauernig

Consequentialism or Virtue Ethics?: A Debate
By Jorge L.A. Garcia and Alastair Norcross

For more information about this series, please visit: https://www.routledge.com/Little-Debates-about-Big-Questions/book-series/LDABQ

Do Numbers Exist?

A Debate about Abstract Objects

Peter van Inwagen and
William Lane Craig

Routledge
Taylor & Francis Group
NEW YORK AND LONDON

Designed cover image: SEAN GLADWELL / Getty Images

First published 2024
by Routledge
605 Third Avenue, New York, NY 10158

and by Routledge
4 Park Square, Milton Park, Abingdon, Oxon, OX14 4RN

Routledge is an imprint of the Taylor & Francis Group, an informa business

ISBN: 978-0-367-44277-4 (hbk)
ISBN: 978-0-367-44276-7 (pbk)
ISBN: 978-1-003-00871-2 (ebk)

DOI: 10.4324/9781003008712

Typeset in Sabon
by codeMantra

Contents

Foreword

Mark Balaguer

It is my pleasure to introduce the spirited debate contained in this book. The debate is about a very old philosophical question—namely, the question of whether there are any such things as *abstract objects* (and, in particular, *numbers*). The parties to the debate are both extremely good philosophers. William Lane Craig is a prominent and important philosopher of religion who also works on issues related to time and abstract objects, and Peter van Inwagen is one of the best metaphysicians of his generation, with seminal works on a number of different topics, including free will and mereology.

In this Foreword, I will do two things. First, I'll introduce the question that the debate is about and explain why it's philosophically interesting and important. And secondly, I'll say a few words about how I think the debate can be simplified.

1. The Problem of Abstract Objects

Philosophers have defined 'abstract object' in numerous ways, but here's a fairly standard way to do it:

> *Definition*: An *abstract object* is an object that's non-physical, non-mental, non-spatiotemporal, and non-causal.

Speaking metaphorically, philosophers often say that abstract objects "exist in Platonic Heaven", but that's just a metaphor; according to the standard platonist view, there is no *place* where abstract objects exist. They *exist*, on this view, but they don't exist in any place. And they're not *mental* things either—e.g., they're not ideas in our heads.

The idea of an abstract object can perhaps be clarified by considering a couple of examples of things that some people have thought are abstract objects. One example is *properties*—e.g., *Redness* and *Roundness*. You might think that Redness—the property of being red—exists independently of red objects like Mars and the Golden Gate Bridge; and if you do, then you might think that it's an abstract object. A second example is *numbers*. Think of the number 3. It doesn't seem to be located anywhere in spacetime—e.g., it's not in Cleveland right now. Of course, there might be three cars parked on some street in Cleveland right now, but those three cars aren't the number 3. What, then, *is* the number 3? Well, some people think that it's an abstract object.

The view that there really are such things as abstract objects is known as *platonism*. This view is extremely controversial. Lots of philosophers—namely, *anti-platonists*—think that there are simply no such things as abstract objects. Notice, however, that if we say that there are no such things as abstract objects, then we have to provide a different theory of what *numbers* are. And it seems that there are only three theories here that are available to anti-platonists. First, they can endorse *physicalism*, i.e., the view that numbers are physical objects. Secondly, they can endorse *psychologism*, i.e., the view that numbers are mental objects, presumably ideas in our head. Or third, anti-platonists can endorse *anti-realism*, i.e., the view that there are simply no such things as numbers.

It turns out, however, that two of these anti-platonist views—namely, *physicalism* and *psychologism*—are not really taken seriously by philosophers of mathematics. As Professor Craig points out in his opening statement in this book, physicalism and psychologism were subjected to devastating criticisms in the late nineteenth century by Gottlob Frege, and since then, they have been almost universally rejected by philosophers of mathematics.

Given that physicalism and psychologism are non-starters, there are really only two views to endorse about numbers, namely (a) the *platonist* view that numbers are abstract objects, and (b) the *anti-realist* view that there are simply no such things as numbers. This is what the debate in this book is about—whether we should endorse platonism or anti-realism about numbers.

Philosophers have debated this question for a long time, but the issue is still very much unsettled. In this book, Professors van Inwagen and Craig weigh in, with the former arguing for the platonist view that numbers do exist (and that they are abstract objects) and the latter arguing for the anti-realist view that there are no such things as numbers.

2. Simplifying the Debate

I just said that there's really only one alternative to the platonist view of numbers, namely, anti-realism. That makes the debate pretty simple. But in his opening statement for this book, Professor Craig runs through numerous *varieties* of anti-realism, and it might seem that to really get a handle on the issues here, we need to understand the differences between these various anti-realist views. But I don't think the differences between the various anti-realist views end up mattering very much; indeed, I think that, at most, there are *merely verbal* differences between the various versions of anti-realism, and I think that, in some cases, there aren't any differences at all.

We can bring this point out by looking at a couple of different cases. I'll start by discussing the difference—or the alleged difference—between the view that Craig calls *fictionalism* and the view that he calls *modal structuralism*.

2.1 Fictionalism vs. Modal Structuralism

On the surface, the difference between these two views seems to be about the question of how our mathematical sentences and theories should be *interpreted*. Consider, e.g., the following two sentences:

[1] '3 is prime.'
[2] 'There are some prime numbers greater than 100.'

Read at face value, sentence [1] seems to say that a certain object (namely, the number 3) has a certain property (namely, the property of being prime); and sentence [2] seems to say that there exist prime numbers of a certain kind. Thus, since we've already decided that numbers couldn't be physical or mental objects, it seems that, on the face-value reading of sentences like [1] and [2], they are claims about abstract objects. We can call this *the platonistic interpretation* of sentences like [1] and [2]. And this, it should be noted, is the *standard* interpretation of these sentences.

Given this, we can define *fictionalism* as the view that (a) the platonistic interpretation of mathematical sentences like [1] and [2] is correct—or to put the point more carefully, sentences like [1] and [2] do *purport* to be about claims about abstract objects—but (b) there are no such things as abstract objects, and so (c) these sentences are simply not true. So on this view, [1] and [2] aren't true for the same reason that, e.g., 'Alice had tea with the Mad Hatter' isn't true—because just as Alice and Wonderland and the Mad Hatter don't really exist, prime numbers like 3 and 101 don't really exist either.

So fictionalism is an *anti-realist* view—i.e., it implies that there are no such things as numbers.

Modal Structuralism is also an anti-realist views. But unlike fictionalists, modal structuralists don't want to claim that sentences like [1] and [2] aren't true. To pull this off, they claim that these sentences can be reinterpreted. In particular, they claim that [1] and [2] can be reinterpreted so that they're equivalent to the following two sentences, respectively:

[1*] 'If the natural-number structure had existed, then 3 would have been prime.'

[2*] 'If the natural-number structure had existed, then there would have been some prime numbers greater than 100.'

Modal structuralists claim that [1*] and [2*] are true, and they point out that we can maintain that these sentences are true *without committing ourselves to the real existence of numbers*. (Compare: we can maintain that the sentence 'If Lewis Carroll's *Alice in Wonderland* had been true, then the Mad Hatter would have existed' is true without committing ourselves to the real existence of the Mad Hatter.)

To get more clear on what's going on here, let me introduce (or invent) two languages—*Literalese* and *Paraphrese*—that are partly defined by the following two stipulations:

(i) In Literalese, the platonistic interpretation of sentences like [1] and [2] is correct; i.e., in Literalese, sentences like [1] and [2] are straightforward claims about abstract objects.

(ii) In Paraphrese, sentence [1] says what [1*] says in English; and sentence [2] says what [2*] says in English; and so on.

Given this, let's ask the following question: What do fictionalists and modal structuralists really disagree about? Both agree that [1] and [2] are *not* true in Literalese. And both agree that [1] and [2] *are* true in Paraphrese. And both agree that we can say that [1] and [2] *are* true in Paraphrese without committing ourselves to the real existence of numbers. So where is the *dis*agreement?

You might have thought that fictionalists and modal structuralists disagreed about whether [1] and [2] are true in *ordinary English*. But, alas, Craig informs us that modal structuralists don't think that [1*] and [2*] provide the *real meanings* of [1] and [2]—i.e., they're happy to admit that ordinary English is equivalent to Literalese, and not Paraphrese. So modal structuralists are happy to admit, with fictionalists, that *in ordinary English*, sentences like [1] and [2] are not true. Moreover, Craig also tells us that modal structuralists don't think that we should change what we're doing in mathematics, i.e., they don't think that we should stop uttering sentences like [1] and [2] and start uttering sentences like [1*] and [2*] instead; so there's no disagreement there either. Now, I imagine that modal structuralists *would* endorse something like the following claim:

> If we *wanted* to, we could stop uttering sentences like [1] and [2] and start uttering sentences like [1*] and [2*] instead—or we could all stop speaking English and start speaking Paraphrese instead—and if we did this, nothing important would be lost; and the fact that nothing would be lost is part of why we shouldn't endorse the real existence of numbers.

But fictionalists can endorse this claim as well, and indeed, I think that most fictionalists *would* endorse this claim—or something essentially equivalent to it. So I just don't think there's any real disagreement between fictionalists and modal structuralists.

Now, if modal structuralists claimed that ordinary English was equivalent to Paraphrese, rather than Literalese, then there *would* be a difference between their view and the fictionalist view. But it would be a *merely verbal* difference that wouldn't matter at all to the philosophical question of whether abstract objects exist. I won't attempt to argue for this here, however, because the issue will reemerge in the next case that I want to discuss—the case that I'll turn to now—and so I'll hold off and argue the point there instead.

2.2 Fictionalism vs. Anti-Quineanism

I turn now to the difference between fictionalism on the one hand and the various *Anti-Quinean* views that Craig discusses on the other. We can define *Anti-Quineanism* as the view that (a) sentences like [1] and [2] should be read literally and at face value, just as the platonistic interpretation tells us to read them; and (b) there are no such things as abstract objects; but despite this, (c) sentences like [1] and [2] are still true.

At first blush, this view might seem very odd. Consider, e.g., the following sentence:

> [Santa] 'Santa Claus brings presents to children on Christmas Eve.'

Since Santa Claus doesn't really exist, it seems that [Santa] isn't really *true*. What actually happens (I'm sorry to report) is that *parents* buy presents for their children, and then they just *pretend* that Santa Claus did it. So, again, [Santa] isn't really true. Or so it seems.

But you might think this is too fast. For you might think there's a *sense* in which [Santa]—or at least *some utterances* of [Santa]—are true. To bring this out, imagine that a Martian came to Earth and was trying to learn our customs and, knowing full well that the whole thing was a pretense, asked us whether Santa Claus brought presents to children on Easter. And imagine that, while correcting the Martian, I uttered [Santa]. You might think there's a sense in which my utterance of [Santa] was true. And you might think that in saying that my utterance was true, we don't commit ourselves to the real existence of Santa Claus.

This is what Anti-Quineans think about sentences like [1] and [2]. They think there's a clear and obvious sense in which [1] and [2] are true—analogous to the sense in which [Santa] is true—but they think that in saying that [1] and [2] are true, we don't commit ourselves to the real existence of prime numbers like 3 and 101.

So that's the Anti-Quinean view. But I want to argue now that there isn't any important difference between this view and fictionalism. One way to argue for this would be to define a language—*Anti-Quinese*—in which sentences like [1] and [2] mean something slightly different from what they mean in the fictionalist's language. This would match the strategy that I used in arguing that there's no important difference between fictionalism and Modal Structuralism. But I think that a better way to proceed is to set things up so that fictionalists and Anti-Quineans *agree* about the meanings of sentences like [1] and [2]—in particular, they both agree that these sentences make straightforward claims about abstract objects—and to then stipulate that the word 'true' means something different in the Anti-Quinean's language than it does in the fictionalist's language. More specifically, we can stipulate that (a) in the fictionalist's language, the word 'true' expresses the concept of *worldly truth*, and (b) in the Anti-Quinean's language, the word 'true' expresses the concept of *deflationary truth*.

There is a lot of overlap between the concepts of worldly truth and deflationary truth. For example, 'Mars is a planet' is both worldly true and deflationary true, and 'Mars is bigger than Jupiter' is neither worldly true nor deflationary true. What, then, is the difference between *worldly truth* and *deflationary truth*? Well, the only differences that will matter for our purposes here concern two specific kinds of sentences—what I'll call *simple object-property sentences* and *existence sentences*—so let me start by defining these two kinds of sentences:

> *Definition*: A *simple object-property sentence* is a sentence whose logical form is '*Fa*'. (In order to understand this definition, you need to know a bit of first-order logic; in particular, you need to know what '*Fa*' means. If you haven't studied first-order logic, then you can use the following slightly less precise definition instead: a *simple object-property sentence* is a sentence that merely says that some specific object has some specific property.) E.g., the following are simple object-property sentences: 'Hilary Clinton is a democrat', 'Mars is a planet', 'Mars is a snowshoe', and so on. And importantly for our purposes, sentence [1]—i.e., '3 is prime'—is a simple object-property sentence.

> *Definition*: An *existence sentence* is a sentence of the form '(∃x)(...x...)'. (In order to understand this definition, you need to know a bit of first-order logic; in particular, you need to know what '(∃x)(...x...)' means. If you haven't studied first-order logic, then you can use the following slightly less precise definition instead: an *existence sentence* is a sentence that merely says that objects of some specific kind exist.) E.g., the following are existence sentences: 'There are some planets', 'God exists', 'There is at least one US Senator from Wisconsin', and so on. And importantly for our purposes, sentence [2]—i.e., 'There are some prime numbers greater than 100'—is an existence sentence.

Given these two definitions, we can say that the only differences between worldly truth and deflationary truth that will matter for our purposes here concern the following four rules:

(i) In order for a simple object-property sentence to be worldly true, the object in question has to *really exist*. For example, 'Mars is a planet' can't be worldly true unless Mars really exists; and '3 is prime' can't be worldly true unless 3 really exists.

(ii) A simple object-property sentence can, in some cases, be deflationary true even if the object in question *doesn't really exist*. For example, 'Mickey Mouse has big ears' and '3 is prime' are deflationary true, and this doesn't require the real existence of Mickey Mouse or the number 3. (You might wonder what *makes it the case* that these two sentences are deflationary true, whereas, e.g., 'Santa Claus has big ears' and '4 is prime' are presumably *not* deflationary true. This is a tricky question; the answer presumably has to do with *our conventions*; but we needn't worry about this here.)

(iii) In order for an existence sentence to be worldly true, objects of the given kind need to really exist. For example, in order for 'There are some cars' to be worldly true, there have to *really exist* some cars; and in order for sentence [2] to be worldly true, there have to *really exist* some prime numbers greater than 100.

(iv) An existence sentence can, in some cases, be deflationary true even if no objects of the relevant kind *really exist*. For example, sentence [2] could be deflationary true even if there don't really exist any prime numbers. (You might wonder what makes it the case that this sentence is deflationary true, whereas, e.g., 'There is a prime number between 20 and 22' and 'There are some cousins of the Mad Hatter in my cupboard' are presumably *not* deflationary true. Once again, the answer presumably has to do with *our conventions*, but again, we needn't worry about this here.)

With all of this in place, I now want to ask the following question: What does the difference between fictionalism and Anti-Quineanism really amount to? They both agree that sentences like [1] and [2] are *not worldly true*. And they agree that [1] and [2] *are* deflationary true. Moreover, they also agree that we can say that

[1] and [2] are deflationary true without committing to the real existence of numbers. (Fictionalists don't *have* to agree with Anti-Quineans about the deflationary truth of sentences like [1] and [2]; they can claim instead that 'deflationary true' doesn't make any sense; I don't think they should say that, but I won't attempt to argue for this here.)

In any event, if fictionalists and Anti-Quineans agree on all of these points, then what do they *dis*agree about? Well, they *seem* to disagree about whether sentences like [1] and [2] are *true*—not whether they're worldly true or deflationary true, but whether they're just *ordinary* true. The reason they disagree about this, it seems, is that they disagree about whether the ordinary-English word 'true' expresses the concept of worldly truth or the concept of deflationary truth. More specifically, it seems that Anti-Quineans think that, in ordinary English, the word 'true' expresses the concept of deflationary truth, whereas fictionalists think that, in ordinary English, the world 'true' expresses the concept of worldly truth. Or, again, so it seems.

But things aren't always what they seem, and I'll suggest in a moment that there is perhaps no disagreement here over the meaning of the ordinary word 'true'. But before I get into this, I want to argue that even if there *is* a disagreement between fictionalists and Anti-Quineans about the meaning of the word 'true' in ordinary English, it's not an *important* disagreement because it's a *merely verbal* disagreement. At bottom, it's a disagreement about the linguistic usage and intentions of ordinary folk concerning the word 'true'—in particular, about whether folk utterances of the word 'true' are best interpreted as expressing the concept *worldly true* or the concept *deflationary true*. But, surely, no one thinks that the linguistic practices of ordinary folk are relevant to the question of whether we should believe that numbers really exist. Surely, no sensible Anti-Quinean thinks that (a) ordinary folk use 'true' to express the concept *deflationary true*, but (b) if they'd just been more Quinean in their usage of 'true' in relation to, e.g., the Santa Claus story, then we would have had good reasons to believe that there really are such things as numbers. So, again, even if fictionalists and Anti-Quineans disagree about the meaning of the word 'true' in ordinary English—and about whether sentences like [1] and [2] are true, in the ordinary sense of the term 'true'—it's not an *important* disagreement; it's not a disagreement that should be relevant to whether we should believe that numbers really exist.

But now notice that there needn't be any disagreement here about what the word 'true' means in ordinary English—or about whether sentences like [1] and [2] are true, in the ordinary sense of the term 'true'. For while Anti-Quineans seem very invested in the idea that the ordinary-language term 'true' expresses the concept *deflationary true*, it's hard to see why fictionalists should care about this issue at all. Fictionalists, I think, should respond to Anti-Quineans by saying something like this:

> I don't have any strong opinion about what the word 'true' means in folk English. If you want to know about folk English, you should ask a linguist, not a philosopher. But more importantly, I don't *care* what 'true' means in folk English—or about whether sentences like [1] and [2] are true in the ordinary sense of the term. We both agree that (a) sentences like [1] and [2] are *not* worldly true; and (b) they *are* deflationary true; and (c) the claim that they're deflationary true doesn't commit us to the real existence of numbers. So we both need to respond to the platonists' arguments for the claim that these sentences are worldly true. But it just doesn't matter whether they're *ordinary-language true*—because it doesn't matter whether the ordinary-language word 'true' expresses the concept of worldly truth or deflationary truth—and I have no opinion on the matter.[1]

So it seems to me that there needn't be any disagreement at all between fictionalists and Anti-Quineans. And in any event, there's no *non-verbal* disagreement between them.

It's also important to note in this connection that *platonists* can respond to Anti-Quineans in much the same way that fictionalists do. In other words, they can say that they agree that sentences like [1] and [2] are deflationary true and that this doesn't commit us to the real existence of numbers. But platonists can then claim that

1. One plausible view here, it seems to me, is that ordinary English is an inconsistent mess—that when ordinary people use the word 'true', they sometimes express the concept of worldly truth and other times express the concept of deflationary truth. But it's also plausible to suppose, I think, that when the chips are down—when we're doing science, and when, as Professor van Inwagen puts it, we're "in the Ontology room"—we use the word 'true' to express the concept of worldly truth.

they think that sentences like [1] and [2] are *also* worldly true—and that this *does* commit us to the real existence of numbers. (Likewise, platonists can respond to modal structuralists by saying, "Yes, of course, [1] and [2] are true in Paraphrese, and this doesn't commit us to the real existence of numbers; but we think that these sentences are *also* true in Literalese—and that this *does* commit us to the real existence of numbers.")

Given that platonists can (and, I think, *should*) respond to Anti-Quineans (and modal structuralists) in this way, it seems to me that the most charitable reading of the platonist's argument for the existence of numbers proceeds as follows:

(I) Mathematical sentences like [1] and [2] are worldly true (in Literalese). But
(II) If sentences like [1] and [2] are worldly true (in Literalese), then numbers exist. Therefore,
(III) Numbers exist.

Seen in this light, premise (II) is utterly trivial. It's trivial in the same way that, e.g., 'All bachelors are unmarried' is trivial—in particular, it follows from what the words in premise (II) mean. (In philosophical lingo, the point here is that premise (II)—or this version of premise (II)—is *analytic*.) This means that anti-realists have to accept premise (II). And so they have to reject premise (I). Which is to say that all anti-realists have to endorse fictionalism in the sense of 'fictionalism' that really matters here.

Here's another way to put this point: Just as the most charitable way to interpret platonists is as claiming that our mathematical sentences and theories are worldly true in Literalese, so too the most charitable way to interpret fictionalists is as claiming that our mathematical sentences and theories are *not* worldly true in Literalese. And given this, we can say that *all* anti-realists are fictionalists.

This is why I think that there's no important difference between fictionalists and other kinds of anti-realists—e.g., modal structuralists and Anti-Quineans. These philosophers all agree on the central claim that premise (I) in the above argument is false—i.e., that sentences like [1] and [2] are *not* worldly true (in Literalese). If there's any disagreement here at all—and, again, there needn't be one—then it's an unimportant verbal disagreement about what certain words and sentences mean in folk English.

2.3 *Upshot*

If these remarks are correct, then the question we need to ask, in order to determine whether we should believe that numbers really exist, is whether platonists have any *argument* for premise (I)—i.e., for the worldly truth of mathematical sentences like [1] and [2]. The answer is that they *do*. Indeed, they have several arguments for this claim. But two of these arguments stand out as the best ones, in my opinion. One of these arguments is essentially the one that Professor van Inwagen gives in this book—that we need to admit that our mathematical sentences and theories are worldly true in order to account for the way that we use mathematics in science and everyday life. And the second platonist argument is that we need to admit that our mathematical sentences and theories are worldly true in order to account for the *objectivity* of mathematics—i.e., for the fact that there's an obvious objective difference between sentences like '3 is prime' on the one hand and sentences like '4 is prime' on the other. How can we account for this objective difference without saying that the former sentences are worldly true and the latter sentences are not worldly true?

The question of whether we should believe that numbers really exist (and, more generally, the question of whether we should believe that abstract objects really exist) comes down, I think, to the question of whether fictionalists—and by 'fictionalists', I now mean those who think that sentences like [1] and [2] are not *worldly* true—can adequately respond to these two arguments for the worldly truth of our mathematical sentences and theories.

But none of this tells us what the *answer* to the question is. Nor will I try to settle that question in this Foreword. Instead, I refer you to the formidable arguments contained in the rest of this book—to Professor van Inwagen's arguments for the platonistic view that numbers *do* exist (and that they are abstract objects) and to Professor Craig's arguments for the anti-realist view that numbers *don't* exist.

Mark Balaguer
Department of Philosophy, California State University, Los Angeles

Acknowledgment

Prof. Craig thanks his research assistant Timothy Bayless for his help in chasing down references and formatting the bibliography.

Opening Statements

Chapter 1

Opening Statement

Peter van Inwagen

Contents

1. Can an Orthodox Christian Consistently Believe in the Existence of Uncreated Abstract Objects?

In an essay called "God and Other Uncreated Things" (Van Inwagen, 2009), I defended the position that **abstract objects**—numbers, **propositions, properties, relations**—exist, and that they can be neither agents nor patients: that is to say, the position that there are abstract objects and that they are incapable of entering into causal relations. And I further argued that, since creation is obviously a causal relation, we must conclude that abstract objects are uncreated. (A conclusion that coheres nicely with another thesis I endorse, namely that abstract objects are, one and all, necessarily existent.)

DOI: 10.4324/9781003008712-2

Clarifications of Some Basic Terms

These terms cannot be defined, either because they are so fundamental to our discourse that they figure in every definition or because any definition of them would be controversial. But there are things that can be said to make their meanings clearer.

Propositions are the items denoted by phrases formed by prefixing declarative sentences with such phrases as 'the proposition that', 'the belief that', 'the opinion that', 'the thesis that', 'the theory that', and 'the hypothesis that'. For example: the proposition that all Greeks are mortal and the belief that everyone has a guardian angel are propositions. Propositions are **expressed by** declarative sentences: the proposition that all Greeks are mortal is expressed by the sentence 'All Greeks are mortal'. Every proposition is either true or false—or if not clearly true or clearly false, at least "in the grey area between" truth and falsity.

Properties are also called attributes, qualities, features, characteristics, and traits. Here are some examples (found on line) of the use of these words:

> Elasticity is a property of rubber; in other words: rubber is elastic.
>
> Among the characteristics unique to animals is gastrulation.
>
> Animals in the phylum Chordata share four key features.

Examples of **relations** are *north of*, *mother of* and *greater than or equal to*; Montreal stands in the relation *north of* to New York; Dorothy Quimby bears the relation *mother of* to Ramona Quimby; π and *e* enter—in that order—into the relation *greater than or equal to*.

Propositions, properties, and relations are **abstract objects.** (I will use 'object' in this very general sense: it is a noun that by definition applies to *everything*. I will sometimes use 'thing' and 'item' in that sense as well.) Mathematical objects like numbers, operations, sequences, spaces, rings, and groups are also abstract objects. However 'abstract object' should be defined, it does not apply to any items of the sorts God calls into being in the first chapter of Genesis—and neither does it apply to electrons, neutron stars, thrones, dominions, principalities, authorities, or God himself. For my best attempt to explain the meaning of 'abstract object', and it is not a very good one, see Van Inwagen (2004).

Objects that are not abstract objects are called **concrete** objects.

But how is this statement to be reconciled with words that I recite with conviction every Sunday?—the words

We believe in one God,
the Father, the Almighty,
maker of heaven and earth,
of all that is, seen and unseen.

(And, a moment later, the words, 'Through him [Christ] all things were made'.) What is more, Holy Scripture says this of Christ:

All things were made through him

(John 1:3)

In him all things were created, in heaven and on earth, visible and invisible, whether thrones or dominions or principalities or authorities—all things were created through him and for him.

(Col 1:16)

I argued that the apparent contradiction was only apparent, owing to the fact that in both the creedal statement (the first sentence of the so-called **Nicene Creed**) and the biblical statements, such phrases as 'all that is' and 'all things' are meant to refer only to things with causal powers—to *real* things (such as thrones, dominions, principalities, and authorities), as opposed to *ideal* things (such as numbers, propositions, attributes, and relations). I pointed out what seemed to me to be the strong analogy of this apparent inconsistency to the apparent inconsistency of the following passage from the *Summa Theologiae* of St Thomas Aquinas (1225–1274),

Therefore, whatever does not imply a contradiction is included among those possible things in respect of which it is said that God is omnipotent. But those things that do imply a contradiction do not fall within the scope of divine omnipotence, since they cannot have the nature of possible things (*ST* I q.25 a.3 *Resp.*).

with Scripture:

For human beings such a thing is impossible, but for God everything is possible.

(Matthew 19:26)

Nothing whatever that God ordains shall be impossible.

(Luke 1:37)

The standard reply on Thomas's behalf is this. The application of the phrases 'everything' and 'nothing whatever' in the Dominical and the angelic statements is restricted: those statements are not to be taken to imply that God can (for example) create a cubical ball, change the past, or break a promise. (As to the last, if they did imply that, Scripture would contradict Scripture: "If we are faithless, [God] remains faithful, for he cannot deny himself" (2 Tim 2:13).) And this is a common enough feature of language. For example, the shopkeeper says at the end of a day of very brisk sales, "What a day! We've sold everything." But she hasn't sold Moscow. She hasn't even sold the cash register, and that's something she could sell, and it's right there in the shop.

Professor Craig, however, has contended that even if my position is not demonstrably inconsistent with Scripture and the Creeds, it is demonstrably inconsistent with the teachings of **the Fathers of the Church.**[1]

The Fathers of the Church

The Fathers of the Church were Christian theologians and preachers of the fourth and fifth centuries. (Or primarily of those centuries. The term 'Father of the Church' has been applied to writers who lived as early as the end of the first century and as late as the eighth century.) It was the Fathers of the Church who systematized Christian theology and who defined Christian theological orthodoxy.

According to Craig, when the Fathers of the Church made statements like 'God is the creator of all things' (and they did), they meant

1. Craig (2016). See particularly Chapter 2, "God: The Sole Ultimate Reality"—and even more particularly, pp. 31–40.

their use of 'all' to be absolutely unrestricted—for they explicitly said that that was how they meant 'all' to be taken. And, further, they made a point of insisting that the "all" that God had created included such items as attributes and the Platonic Forms (included them conditionally, that is: *if* there are such things, they are created things).[2] But if the Fathers said these things, does it follow that the position I defended in "God and Other Uncreated Things" is inconsistent with their teachings? It is far from obvious to me that it does follow. I will attempt to explain why I say this. I begin by telling a story, a sort of parable.

> A certain liberal theologian of our time is a fervent champion of the sacramental validity of same-sex marriages. Oddly enough, his principal argument for their validity rests on an appeal to the Fathers. Several of the Fathers—writing in opposition to the heretical doctrine that only marriage ceremonies performed according to Christian rites were sacramentally valid marriages—affirmed that *all* marriages that were recognized as valid by the laws of a particular jurisdiction and satisfied two conditions were sacraments and true marriages in the sight of God. And the two conditions were: First, that neither party had once married a person who was still living; secondly, that the vows of both parties had been said with a "right intention." (As to the meaning of "right intention," an example of an intention that is not right must suffice. If the woman had said her vows only because her family had threatened to punish her if she refused to say them, then she did not say them with a right intention.) Our liberal theologian enlists these Fathers in

2. The reader will perhaps be puzzled by the fact that, although the title of this book is *Do Numbers Exist?: A Debate About Abstract Objects*, I hardly mention numbers in this, the long opening section of my contribution to this book. The next part of my contribution will be devoted to numbers, but (very nearly) the only abstract objects I consider in the present section are properties. In my defense, I can say only that, first, I did not choose the title, and, secondly I consider the question, "Are there such things as properties?" to be the fundamental question concerning the ontology of abstract objects. I would continue my defense by saying that *if* there are properties, then *of course* there are numbers—in the sense that there are, well, any *number* of ways to "identify" numbers with certain propositions or certain properties (none of them the one "right" way). For example, the natural number 2 might be identified with the proposition there are at least two things (that is, that there is a thing and another thing) and the number 3 with the proposition that there are at least three things—and so on. But, of course, it might as well be identified with any of vastly many other propositions or with any of vastly many properties.

> his cause—for in many jurisdictions of the present day, same-sex marriages are recognized by law.

So goes my parable. Now let us ask, when our imaginary Fathers contended that all marriages that were valid according to the laws of any jurisdiction (and which satisfied their two further conditions) were sacramentally valid, did they mean their **quantifiers** to be unrestricted? I will so stipulate. But then I would ask, what if they were asked to imagine a state whose laws permitted the marriage of two people of the same sex—whose laws implied that two men or two women could be legally married in exactly the same sense of 'legally married' as the sense in which a man and a woman could be legally married? I find it hard to suppose that they would continue to say that *all* marriages that were valid according to the laws of any jurisdiction (and which satisfied their other two conditions) were sacramentally valid. I would expect them to say that they had never even *thought* of the possibility of a state whose laws would permit marriage between two people of the same sex—and that *of course* such marriages-according-to-the-law would not be sacramentally valid.

Now one might quarrel with the details of this example. Perhaps (it might be objected) an orthodox theologian of the Patristic Age to whom this possibility was mooted would continue to endorse the unqualified 'all' statement on the ground that same-sex marriages were not even *legally* possible, since the very *concept* of marriage implied that only a man and a woman could be married to each other. But—surely?—the lesson I intended it to teach stands even if some feature of the example that unfits it for teaching that lesson. And the lesson is that one might endorse a certain 'all'-statement, mean it to hold without any possible exception, and, nevertheless, *would* admit that there were possible exceptions to it *if* certain possibilities one had not thought of were brought to one's attention.

I am happy to concede that if any of the Fathers whom Craig cites could have slept to the present day—not in the sleep of death, but in some such way as people sleep in Arthurian legend—and had been awakened, and had learned English, and Craig had then said to him, "This fellow van Inwagen claims to be a Christian, and yet he thinks that properties (attributes, qualities, characteristics, features, . . .) are necessarily existent and are, moreover, uncreated," that learned revenant would very likely respond by saying something like, "Then he is a heretic." But if there were such an awakened Father, and if he said that, I shouldn't regard his saying it as a demonstration that my views were inconsistent with Patristic orthodoxy. I say that because I do not think that he (the awakened Father) would have any idea, any idea at all, of what I meant by 'property' or any inkling of my reasons for thinking that there were such things as the things I call 'properties'. Similarly, if he were informed there were now philosophers who studied the behavior of natural bodies—they're called "physicists," he's told—who would say that a boulder sitting on the top of a hill was capable of "doing work," he would suppose that those savants were mad. Slaves and horses and oxen are the sorts of thing that do work, he would protest, not inanimate objects. And if 'work' meant always and only what the Greek word *ergon* meant, he would be dead right. But 'work' does not mean the same thing in, e.g., the two statements 'Much work is still to be done before the harvest is safely gathered in' and 'The work-energy theorem is a fundamental principle of thermodynamics'. The two meanings are not entirely unrelated—as are the meanings of the two words 'bank' and 'bank' (two words, despite the fact that they are spelled and pronounced the same) when we use them to talk of financial institutions, on the one hand, and the margins of rivers, on the other—but neither are they identical.

I obviously cannot present a proper argument for the conclusion that the concept I express by the word 'property' (or any concept even remotely resembling it) was wholly unknown to the Fathers. I cannot even present a proper argument for the conclusion that they never explicitly mentioned that concept. (They might have had that concept without ever mentioning it. For all I know, none of the Fathers ever explicitly mentioned the concept "tacking against the wind" or the concept "dactylic hexameter," but I'm reasonably confident that they all *had* those two concepts.) I could not do the latter even if I were far more learned than I am, for my space is limited. But I will give one example in aid of this contention. Consider this passage:

> ... without [body, colour] has no existence (not as being part of it, but as an attendant property co-existing with it, united and blended, just as it is natural for fire to be yellow and the ether dark blue)...

Craig (2016, p. 35) has quoted these words from the *Plea for the Christians* of the second-century Christian philosopher Athenagoras of Athens. (In the remainder of this essay, page citations without reference to a work are to Craig, 2016.) I would certainly say that if colors could not exist apart from bodies but co-existed with them, not as parts of them but united and blended with them, then it would indeed be heretical to say that colors existed and God had not created them. But my colors, that is, the properties of physical things I call 'colors', are nothing at all like Athenagoras' colors (i.e., their nature is nothing at all like what Athenagoras *supposed* the nature of colors to be). Within my metaphysical system, if I may be supposed to have so grand a thing, it makes no more sense to say that the color green is united and blended with a green shamrock than it does to say that the number four is united and blended with a four-leaved shamrock.

I further contend that, whatever metaphysical concepts may have been available to the Fathers, my concept of "property" was quite unknown to the pre-Socratic philosophers, to Plato and Aristotle, and to all the later philosophers of the ancient Mediterranean world. (I will expand on that statement in the section that follows.) If I am wrong about this, it would suffice to show it if one could point to a passage in which one of those philosophers mentions or expresses this concept.

But let us return to our Father who has slept and been awakened in the present day. I do not find what he would say about my views if his knowledge of them were based entirely on the statement, "Van Inwagen thinks that properties are necessarily existent and uncreated" an interesting question. What I *do* find an interesting question is this: Suppose he should be willing to enter a couple of months of dialectical exchange with me about the contents of the various metaphysical categories; would he finally say, "Ah, now I see what you mean by 'property'. It's wholly unlike anything I had ever thought of. But you're still a heretic if you say they're uncreated"? (Or other words to that effect.) Or would he say, "Ah, now I see what you mean by 'property'. It's wholly unlike anything I had ever thought of. It's perfectly all right for a Christian who believes in such things to say that they are uncreated. Go in peace, to love and serve the Lord."? (Or something to much the same purpose.) Or would he say something else, something that hasn't occurred to me? I confess that I have no idea what the awakened Father would say in those circumstances. (I should not even be willing to predict that *two* such awakened Fathers would both say the same thing.) I am, however, confident that, whatever he said, it would include something along the lines of, "It's wholly unlike anything I had ever thought of."

In Section 1, the problem of the creation of abstract objects is introduced: does the statement "God created everything" imply that God created numbers and qualities? This question is compared with the question, does the statement "God can do anything" imply that God is able to make a cubical ball or end his own existence? The almost universally accepted answer to the latter question is No. Van Inwagen contends that the answer to the former question is also No, and that the meaning of 'God created everything' is that God created everything that can be either a cause or an effect—and abstract objects can be neither. Craig has contended that this position is inconsistent with the testimony of the Fathers of the Church, since they affirmed that 'God created everything' has no exceptions, and have particularly affirmed that Platonic "Forms" and Aristotelian "universals" are not exceptions. Van Inwagen, however, suggests that the entities *he* calls abstract objects were unknown to the philosophers of the ancient world. He further suggests that although someone may say that something or other is true of everything, and state vehemently that there can be no exceptions to this generalization, that person might think otherwise if confronted with cases he or she had been unable to conceive of.

2. Lightweight Platonism: Things That Can Be Said of Things

But what is it that I mean by 'property'—and what makes me so sure that this meaning is "quite different from" anything the Fathers had ever thought of? And what is the ground of my confidence that my concept "property" was unknown to Plato and Aristotle?

We must, of course, concede that Plato and Aristotle and the Fathers of the Church did not speak English or any other present-day language and thus did not use the word 'property'. And any answers to the questions, what classical (or Patristic) Greek word corresponds to the English word 'property'? and what Latin word corresponds to the English word 'property'? would be controversial.

Greek, Latin, and the Fathers of the Church

One classification of Fathers of the Church is by the languages they wrote in: thus, we have the Latin Fathers and the Greek Fathers. The Greek of the Greek Fathers is called Patristic Greek (and the era in which they lived and wrote is called the Patristic Era). The word 'Patristic' comes from the Latin word *pater* (father). Patristic Greek is not terribly different from the "classical" Greek of Plato and Aristotle.

One could, however, provide reasonably uncontroversial translations of such words as 'roundness' and 'solidity' and 'swiftness' into those languages. Let us use 'roundness' as our example of such a word. Ancient ideas about what 'roundness' might refer to were of two general kinds, one due to Plato and the other to Aristotle.

For Plato, horses and human beings and battleships and all other tangible, visible things are like dead leaves in a high wind: present at one moment and gone the next, gone like the snows of yesteryear. There are, however, other things than tangible, visible things and some of these, at least, are *not* like leaves in the wind. There are things, Plato holds, that stand firm and immovable, untouched by time. Plato calls them "the forms" or "the ideas."

Among the transient, visible things are the round things, the things whose surfaces are approximately spherical—the star Sirius and the moon and (let us say) a tennis ball that is before me as I write. (In the strict and geometrical senses of the words 'ball' and 'sphere', a ball is a solid object, and a sphere is the *surface* of a ball. But I'll sometimes follow common usage and speak of stars and moons and tennis balls as spheres.) What all spheres have in common, and share with nothing else, is sphericality—which I'll call by the more homely name 'roundness'. And there is only one roundness: the roundness of the star, the roundness of the moon, and the roundness of the tennis ball are one and the same thing: a form or idea called "the sphere." (Think of the phrase 'the sphere' on the model of the phrase 'the lion' as it is used in the sentence 'The lion is a well-muscled cat with a long body, large head, and short legs'.) The sphere would exist even if there were no spheres—that is no visible, tangible spherical things.

The unchanging and eternal forms, Plato tells us, are the most real things, the only fully real things. The form "the sphere" is more real than the abstract spheres geometers are referring to when they say things like, "If the area of a sphere is four times the area of another sphere, the volume the larger sphere encloses is eight times the volume the smaller encloses." And the ideal objects spoken of in that statement, Plato holds, are more real than our tennis ball—just as that tennis ball is more real that its shadow or its reflection in a pool. (The physical ball is something like a shadow or reflection of the balls of the geometry texts, and those balls are in the same way shadows or reflections of the form "the sphere.")

According to Plato, the form or idea of "the sphere" is, in a sense appropriate to its exalted metaphysical status, a *cause*: it is the cause or ground of the existence of the many ideal spheres treated in geometry, and all the truths about them; it is the cause or ground of the possibility of there being tangible, material things that, like this tennis ball, are at least approximately spherical for some brief segment of eternity. It is the cause or ground of the fact that it is possible for beings like you and me to entertain thoughts like 'The moon and this tennis ball have the same shape' and 'The intersection of a plane and a sphere is always either a single point or a circle'. It is the forms or ideas that account for such constancy as there is amidst the flux and impermanence of the sensible world, and it is the forms that account for the fact that our thoughts can be about matters that go beyond what is immediately present to our senses.

Plato's student Aristotle would have none of this—or almost none of it. He affirmed the existence of entities like "roundness" and "solidity" and "swiftness"[3] but condemned Plato for having separated the qualities (attributes, properties, features, characteristics, etc.) of things from the things. It is often said that Plato's Forms exist "in the Platonic Heaven"—with the consequence that "the Round" is (in some nuanced, metaphysical way) far removed from the planets and billiard balls and ball bearings that are round only because (in some nuanced, metaphysical sense) they "participate in" it. No, says Aristotle, the being of these entities, these "universals," is not separate from the being of the things that fall under them. (Aristotle was the inventor of the term 'universal'—that is to

3. What Aristotle's believed about abstract qualities like roundness is a matter of dispute. The view I attribute to Aristotle would, however, be affirmed by most Aristotle scholars.

say, he coined the Greek word *kath'holou*, and the Latin word *universale* was later invented to be the Latin equivalent of *kath'holou*. Plato, in a discussion of virtue, had spoken of "virtue *kata holou*," virtue *taken as a whole*, that is virtue considered in itself and apart from any of the things that are virtuous. Aristotle made the noun *kath'holou* from this phrase.) Rather, roundness "inheres in" round things, solidity inheres in solid things, and swiftness inheres in swift things. And this means that the being or existence of a quality depends on the existence of things of which it is a quality: roundness can exist only *in* round things, solidity only *in* solid things, and swiftness only *in* swift things.

We will call Plato's metaphysic of things like solidity and swiftness and roundness **Platonism**. And we will use the term **Aristotelianism** to designate any metaphysic of roundness *et al.*—of universals—that holds that roundness can exist only in round things, solidity only in solid things, swiftness only in swift things, and so on.

There are philosophers who say "A plague o' both your houses" to Plato and to Aristotle. These philosophers are generally called **nominalists**. Nominalists simply deny that there is any such thing as solidity, swiftness, or roundness. "Yes," say the nominalists, "the ball is round. Who could deny it? But there is no such thing as the roundness of the ball—and there is no such thing as the solidity of the solid oak table or the swiftness of swift Achilles."

Platonism, Aristotelianism, and Nominalism

Platonism: Words like 'solidity', 'swiftness', and 'roundness' refer to eternal, unchanging objects. The phrases 'The roundness of Sirius', 'the roundness of the moon' and 'the roundness of the ball' refer to the same thing, a thing whose existence is independent of the existence of Sirius and the moon and the ball, a thing that would exist even if nothing at all were round.

Aristotelianism: The word 'solidity' refers to something that exists only in solid things; the word 'swiftness' refers to something that exists only in swift things; the word 'roundness' refers to something that exists only in round things. If nothing were solid, there would be no solidity; if nothing were swift there would be no swiftness; if nothing were round there would be no roundness.

Nominialism: Words like 'solidity', 'swiftness', and 'roundness' refer to nothing at all. (For all that, they may be useful and have a role to play in our discourse: compare 'the average American family'). The name is taken from the name of a late-Medieval philosophical position (from Latin *nomina*, 'names'). The present definition of 'nominalism' should not be regarded as statement of the "nominalism" of the Middle Ages.)

I affirm, with Craig and the Fathers, that if either Platonism or Aristotelianism is right, the proposition

> Solidity, swiftness, and roundness exist and were not created by God

is incompatible with Christianity, and indeed with theism. If Aristotle is right, those three items stand in causal relations to individual things. (The presence of solidity in the table *prevents* one from putting one's hand through it, and one can *observe* the swiftness of Achilles and the roundness of the ball.) If Aristotle is right, therefore, the three items are undeniably parts or components or constituents of what Christians call (and all theists call) Creation. And if Plato were right—if there were eternal, unchanging things upon which the continuity and intelligibility of the world of sensible things depended—then, if those items were uncreated, matters would be even worse. If Plato's Forms existed, and if they were uncreated, they would be, so to speak, impersonal rivals of God.

I also affirm, however, that neither Platonism nor Aristotelianism *is* right. (And not because nominalism is right.) The position that I say is right is in agreement with Platonism on two points:

(i) Nominalists are wrong to say that words like 'solidity', 'swiftness', and 'roundness' designate nothing. Rather, they designate **universals**—for 'roundness' is universal to round things (and solidity is universal to solid things, and swiftness to swift things ...).

(ii) Aristotle was wrong to say that universals have being only in the things to which they apply. Solidity and swiftness and roundness and the like are (to use the medieval terminology) **universals *ante res***, not **universals *in rebus***. (That is, they are "universals *prior to things*" and not "universals *in things*." '*Res*' and '*rebus*' are different forms of the Latin word for 'things'; a feature of Latin grammar that has no English counterpart requires '*res*' after '*ante*' and '*rebus*' after '*in*'.). That is to say, their being is independent of the being of such things as they may apply to. For Plato (and for me) it is an axiom that solidity would exist if the universe were entirely gaseous, that swiftness would exist in a world in which there was no motion of any kind, and that there would be such a thing as roundness if every physical thing were cubical. (We should note that sentences like 'Solidity does not exist' are ambiguous. Consider, for example, 'Honesty does not exist'—one of the sentences that is like that sentence. If someone said, "Honesty does not exist" in an ordinary conversation, that person would be making the cynical statement that no one is honest. But 'Honesty does not exist' can also be used to express a quite different proposition, a purely metaphysical proposition, a proposition entailed by nominalism.) We agree, Plato and I, that not only would roundness exist if nothing were round, but it would exist *no matter what*. The Aristotelians say that there would be no roundness if there were nothing round. *I* say roundness would exist even if there were nothing material at all.

Universals; Universals *ante res*; Universals *in rebus*

Universals are abstract qualities like solidity, swiftness, and roundness. Solidity is a quality of all solid things (and a quality of nothing else). Swiftness is a quality of all swift things (and of nothing else). Roundness is a quality of all round things (and of nothing else).

A **universal *ante res*** is a universal whose being is independent of the being of "things"—that is, independent of the being of the entities it is capable of being present in. If roundness (for example) is a universal *ante res*, it would exist even if there were nothing round.

A **universal *in rebus*** is a universal whose being is somehow parasitic on the being of the things it is present in. If roundness is a universal *in rebus*, it can no more exist independently of "things" than a surface or a hole can exist independently of solid bodies.

If I agree with Plato on these two points, I disagree with him on many other, and perhaps more important, points. For one thing, I deny that universal "roundness" is *more real* than a tennis ball. In fact, I deny that anything is more real than anything. I also deny that universals are the *causes* or *grounds* of any aspect of the world of tangible, visible things. Or, rather, I do not so much deny these theses as deny that that *are* any such theses. In my view, sentences like 'Roundness is more real than round balls' and 'The universal "roundness" is the cause or ground of the existence of the ideal spheres of solid geometry, and all the truths about them' and 'The universal "roundness" is the cause or ground of the possibility of there being tangible, material things that are at least approximately spherical' are simply meaningless. But I won't press this point. The point I will press is that my "roundness" is a much wispier thing than Plato's "the form of the round."

My former student Dr. Kenny Boyce has called me a "lightweight" Platonist, and I have adopted the term myself—with the added cautionary device of spelling 'platonism' with a lower-case 'p'. (Let 'Platonism' with a capital be reserved for the philosophy of the man who actually bore the name 'Plato'.) And, indeed, most present-day philosophers to whom the p-word is applied are much more "lightweight" than Plato. Most present-day philosophers would probably agree with what is said in the following passage from the article "Abstract Objects" in *The Stanford Encyclopedia of Philosophy*:[4]

> ...it is therefore important to remember that the use of the terms platonist (for those who affirm the existence of abstract objects) and nominalist(for those who deny existence) is somewhat lamentable, since these words have established senses in the history of philosophy. These terms stood for positions that have little to do with the modern notion of an abstract object. Modern platonists (with a small 'p') need not accept any of the distinctive metaphysical and epistemological doctrines of Plato.

I will henceforth in this book refer to my position as **lightweight platonism.** (But be warned: Craig makes some use of the term 'lightweight Platonism'—he uses the capital—, but what he means by 'lightweight Platonism' is not what I mean by 'lightweight

4. Falguera, Martínez-Vidal, and Rosen (2021), Section 2.1.

platonism'.[5] In my view, the philosophers whom Craig classifies as lightweight Platonists should not be classified as Platonists of any sort.)

A Statement of Lightweight Platonism—Part I

There exist (in the only sense of 'exist' there is) propositions, properties, and relations. However 'abstract object' is to be defined, every abstract object is a proposition, a property, or a relation. (And, therefore, any mathematical object—a complex number, a vector space, an Abelian group—belongs to one of these three categories.) All abstract objects are necessarily existent, and are, moreover, essentially without causal powers: that is, they are incapable of entering causal relations (in any *possible* sense of 'causal relation').

Although Platonists (including lightweight platonists) affirm, and their nominalist opponents deny, the existence of propositions and relations, the "universals" Platonists and nominalists are most often to be found disputing about are properties (qualities, attributes, etc.). In the remainder of this discussion, I will concentrate on properties, although what I say could be fairly easily applied to propositions and relations.

These are the questions before us:

> What do I mean by 'property'—and what makes me so sure that this meaning is "quite different from" anything the Fathers had ever thought of? And what is the ground of my confidence that my concept "property" was unknown to Plato and Aristotle?

In this section, I will attempt to answer them. My answer will take this form: I will present my **ontology** of properties (of the things *I* call properties; some may think that that's a misleading thing to call them) and leave it to the reader to judge whether these things were indeed unknown to the Fathers and to Plato and Aristotle.

5. See Section 1 of Craig's opening statement in this volume, particularly pp. 81–82.

The word **ontology** (from Greek *on*, being) has three closely related meanings. In two of its senses it denotes a certain study or discipline. What we may call "classical ontology" is devoted to the question of being: What is being? What is it for a thing to *be*? And what we may call "analytical ontology" is devoted to the question? What is there? ('Are there properties, numbers, or any abstract objects of any sort?' is a question that is addressed in analytical ontology.) The third sense of 'ontology' is 'a position in analytical ontology'. When one is using the word in this third sense, one can speak of "ontolog*ies*" and "*an* onotolgy." The word has this sense in statements like, "I am a nominalist; there are no abstract objects in *my* ontology" or "I am a platonist; my ontology includes both attributes and mathematical objects."

I begin by spelling out the sense in which the things I call properties were "unknown" to Plato and Aristotle and the Fathers—for there is one sense in which they were very well known to them indeed. Here is an imaginary exchange between two theologians of the Patristic Age. (Imaginary, but it cannot be doubted that many real conversations in those days included exchanges very much like it.)

> "Have you heard what Damian of Cyrene said about Dismas of Ancyra?"
>
> "I have indeed. I'm told that Dismas is furious. But Damian has never been known for his originality. Erechtheus of Pergamum said the same thing about Philologus of Phrygia years ago."

It will in due course become evident to the reader that two phrases in this exchange—'what Damian of Cyrene said about Dismas of Ancyra' and 'the same thing'—denote what I call a "property." (Or would if the story were historical—and, of course, they denote one of these things only if there are such things.) There is therefore a sense in which philosophers and theologians of the classical and Patristic periods had the concept I express by the word 'property': it is implicit in their thought and discourse, but, as one might say, they did not know that they had it. Or, again, although they had it, they did not have it in a way that allowed them to bring it before their minds to discuss it or make use of it in their thinking. It would not have come to the mind of any of the Fathers if he were attempting to list all possible answers to questions like, "What is a *poiótes*" or "What is the nature of a *qualitas*?" (If someone were, for some reason to choose to translate a present-day work in English on the metaphysics of properties into classical or Patristic Greek or into classical or medieval Latin, those words—the first is Greek, the second is Latin—would be plausible translations of any of the English words 'property', 'attribute', and 'quality'.)

Examples may help me to convey the idea of a concept that is implicit in the language of a certain community but is not available to its members in a form that they can consider or reflect on. I will offer two.

The first is the concept "person." The twentieth-century English philosopher P.T. Geach has written (Geach, 1977, pp. 75–76):

> The concept of a person, which we find so familiar in its application to human beings, cannot be clearly and sharply expressed by any word in the vocabulary of Plato or Aristotle; it was wrought with the hammer and anvil of theological disputes about the Trinity and the Person of Christ ... The familiar concept of a person finds linguistic expression not only in the use of a noun for 'person' but also in the use of the personal pronouns 'I, you, he'.

If Geach is right, Plato and Aristotle *had* the concept of a person (it was implicit in their use of pronouns—and, I would add, personal pronouns are not the only pronouns in which the concept of a person "finds linguistic expression": in English, for example, 'who' and 'someone' are reserved for persons). But pre-Christian philosophers, not having a word or phrase that expressed the concept "person" could not, as the late-fifth- and early-sixth-century Christian philosopher Boethius could, address the question 'What is a person?'

My second example is the concept "time." Although much of what the twentieth-century American linguist Benjamin Lee Whorf said about the language of the Hopi people of northern Arizona was wrong (so most linguists of the present day say, at any rate), he was, I understand, right to say that there is no abstract noun or noun-phrase in Hopi that expresses the concept "time." But it would be wrong to conclude from that fact that the pre-European-contact Hopi did not *have* the concept "time"—for one can of course say things like, "The boy *is* promising, but he *is not yet* a skilled hunter" and "She married him *before he became* crippled" and "I *will meet* you there *at moonrise*" in Hopi. That is, the concept of time is implicit in the linguistic practices of Hopi speakers. It would, however, be right to conclude that the philosophical question 'What is time?' cannot be raised in Hopi—and that it is therefore not a question that a philosopher who spoke only Hopi could ask, much less undertake to answer.

When I say that the things I call 'properties' were unknown to pagan Greek philosophers and to first-millennial Christian theologians and philosophers, therefore, I do not mean that they did not refer to and generalize about them in their daily speech and thought; I mean that their conceptual resources were not such as to enable them to think *about* them.

But what *are* they?

I offer now an introduction to lightweight platonism. It will concentrate on properties (the objects *I* call 'properties'; whether this is an appropriate name for them is in my view a purely verbal question). Its primary purpose is not to defend lightweight platonism but rather to convince the reader that the concept "property" that figures in lightweight platonism is a concept that the Fathers of the Church did not have (in the sense of 'did not have' I attempted to spell out in the preceding section).

It would seem that we all believe that there are things that people say. Consider, for example, the following sentences, each of which is certainly a sentence that might be used to say something:

Paris is the capital of France

The capital city of France is Paris

Paris is the French capital

Paris est la capitale de la France

Parigi è la capitale della Francia

Paris ist die Hauptstadt von Frankreich.

If someone who understands any of these sentences speaks or utters that sentence in, oh, let's say, assertion-friendly circumstances (the speaker is addressing someone whom the speaker believes understands the sentence; the speaker is not an actor speaking a line in a play; the speaker is not engaged in reading sentences aloud from a sequence of cards successively displayed by a physician who is trying to determine whether the speaker is concussed . . .) that person *says something*—that is, asserts something. And if Alice has said something by uttering one of these sentences and Bertram has said something by uttering another of them, then Alice and Bertram have *said the same thing in different words.*

If there are things that people say, there are things that people might have said but never have said and never will say. It is, for example, extremely unlikely that anyone ever has said or ever will say that every building in San Francisco is a veterinary hospital; but that is something that it would be *possible* to say. There are, therefore, *things that have not been said but could be said* or *unsaid things that it would be possible to say.* (Are there not, as the proverbial wisdom has it, things that are better left unsaid?) And there are vastly many of them—so many that only a vanishingly small proportion of them will ever actually be said.

I will call things that it is possible to say *propositions.* (But if you don't want to use that word for them, feel free to call them by some other name—I'm with Juliet: What's in a name?)

Now one important property of propositions is that they are either true or false (or perhaps in the gray area between truth and falsity). The concept of a proposition "involves" the concepts of truth and falsity in this sense: A person cannot grasp or understand the concept "proposition" unless that person also grasps the concepts "truth" and "falsity." The concept of a proposition is therefore a *truth-involving concept*—and, we may say, propositions are truth-involving objects and the kind "proposition" is a truth-involving kind. In fact, having introduced propositions as things that someone could say, I will now confess that this was a bit of a cheat—an expositional trick to aid in explaining what I mean by 'proposition'. For what I *really* mean by 'proposition' is 'something that is either true or false' and it seems plausible to suppose that there are both truths and falsehoods that cannot be said—that is, asserted. There is, for example, for each real number the proposition that it is greater than 2, and almost none of *those* propositions can be said—not at any rate by any created being. Having noted this, I'll for the most part stick with my expositional trick in what follows.

If "proposition" is a truth-involving concept, however, it is not the only one, and propositions are not the only truth-involving objects. For, in addition to things that can be said, there are things that can be said *of* things, and these things are true or false *of* things. For example: "that it is a national capital," "that it is cubical," and "that it has proper parts." If Janet says, "Paris is a national capital" and Étienne says, "*Washington est une capitale nationale*," then, it would seem, the two speakers have both said something *and* have both said something *of* something. And, although what the one *said* was not what the other *said*, what the one said *of* something is what the other said *of* something. (*Different* "somethings," to be sure, for Janet said it of Paris and Étienne said it of Washington.) That is to say, *that it is a national capital* (an item that is also called '*that it is the capital of a nation*' and '*qu'il/elle est une capitale nationale*') is something that can be said of things. And the concept of a thing that can be said of things is a truth-involving concept—for *that it is a national capital* can be said *truly* of some things (Paris and Washington and Rome) and only *falsely* of others (Florence, the Taj Mahal, you and I, the cube root of 2 . . .). And it is essential to one's grasping the concept "thing that can be said of things" that one be aware that some things that can be said of things can be said truly of some things and only falsely of others.

I will call things that can be said of things 'properties'. But if you don't like calling them that, feel free to substitute 'thing that can be said of things' or any other word or phrase you like for 'property' in the sequel. (Again, I'm with Juliet.)

I say that a thing *x has* the property *y* (or that *y belongs to x* or that *y is a property of x*) just in the case that *y* is true of *x*. (I much prefer sturdy, quotidian usages like 'has' and 'belongs to' and 'is a property of' to highfalutin words like 'exemplifies' and 'instantiates'. Consider this "real" sentence, which I found on line: 'Plaster has the unfortunate property of gradually dissolving when in contact with water'. No one would say that plaster "exemplifies" or "instantiates" the unfortunate property of gradually dissolving when in contact with water.) Thus, in my usage, 'that it is white' and 'whiteness' are two names of the same object, and the Taj Mahal has (instantiates, exemplifies) whiteness just in the case that *that it is white* is true of (is something that can be said truly of) the Taj Mahal.

What I have said implies that a "property" can exist if nothing has it—for obviously, there are things that can be said of things that can't be said truly of anything. (That it is a perpetual-motion machine, for example.) But there is more: a property not only *can* exist if nothing has it, it *must* exist whether anything has it or not.

That is, every property is *necessarily* existent. Suppose, for example, that someone suggests that the property *that it is cubical* might not have existed. If that's right, there's a possible world in which there's no such thing to be said of something as *that it is cubical*. I don't mean a world in which there's nothing that *has* that property—that is, a world in which there are no cubes. Of *course* it's possible for there to be no cubes. I mean a world in which there's no such property for things to have. In such a world, the existence of cubes wouldn't even be a possibility. In such a world an artisan couldn't say, "I've thought of a very elegant shape-property. It's the shape-property that would be exemplified by something if and only if it was a regular convex polyhedron with six faces, all of them squares (and with eight vertices and twelve edges). I'll make something with that property next week." For in that world, by definition, there's nothing for the phrase 'the shape-property that would be exemplified by something if and only if it was a regular convex polyhedron with six faces, all of them squares' to refer to. In such a world, this actual state of affairs in which we find ourselves would not even exist as a *possibility*—for, in this actual state of affairs, there are cubes. And the proposition 'It is possible for there to be cubes' can't be true unless *that it is cubical* is a possible shape-property. And it can't be a possible shape-property if it's not there at all. So if *that it is a cube* fails to exist in some possible world, this state of affairs we find ourselves in not only might not have been *actual* (practically everyone but Spinoza concedes *that*), it might not even have been *possible*. If *that it is a cube* could have failed to exist, then this actual state of affairs could have failed to be possible. But I affirm as a metaphysical axiom that if a state of affairs is actual, that it is a *possible* state of affairs is a necessary truth: if a state of affairs is actual, then it's necessarily possible.

Or think of matters this way. According to orthodox Christian theology, God might have chosen not to create anything: he had absolute free will in the matter of creation. Well, suppose God had chosen that option. Since God is obviously not himself cubical, there would then have been nothing cubical. Still—let us shift to the indicative mood—God knows about cubical things as possibilities. That is, he knows that he has the power to create a universe some of whose constituent objects are cubes. So he must know about the *property* "being cubical"—or *that it is a cube*. He *must* contemplate that property and form the consequent judgment 'I could have created things that had that property'—for the simple reason that if he did not contemplate that shape-property and form that judgment, he would not be fully aware of every aspect of his power. And, of course, if God contemplates a property, there *is* a property that he is contemplating.

A Statement of Lightweight Platonism—Part II

Properties (qualities, attributes, characteristics, features, traits . . .) are things that can be said of or about things. One of the things you can say about the Washington Monument is that it is white. And you can say it—the very same thing you can say about the Washington Monument—about the Taj Mahal, as well. We could call this thing *that it is white*. 'Whiteness' and 'the property of being white' and '*that it is white*' are three names for the same thing: the thing we *ascribe to* or *predicate of* an object when we say of it that it is white.

An object **has** a property when that property is true of that object—that is, when one says that property of that object, what one says is true. Thus, the White House has whiteness because one who says *that it is white* of the White House speaks the truth; the Brooklyn Bridge does not have whiteness because *that it is white* is not true of the Brooklyn Bridge.

Properties can exist if nothing has them. In fact, they *must* exist whether anything has them or not.

Now let us ask: did God *create* the property he is contemplating—"being cubical" or *that it is a cube*? Well, obviously not if he has not created anything. And we are imagining a state of affairs in which he has not created anything. But let us set that difficult case to one side and ask whether he has *actually* created that property in *our* world. (Note that I'm not asking whether he created *cubes*. Of course he did—all of them.) Well, it's certainly hard to conceive of God or any being *deciding* whether to create a certain property. If he decides not to create it, then what is it he's decided not to create? That question is not parallel to this question: Suppose God is deciding whether to create cubes and decides not to. Then there are never any cubes. What is it he's decided not to create? In that case, he has contemplated the property *that it is a cube* and has decided not to create anything of which that property would be true. "Well, then, when he's deciding whether to create *that it is a cube*, he's contemplating the property, *that it is the property 'that it is a cube,'* and is deciding whether to create anything of which *that it is the property 'that it is a cube'* is true." I find it impossible to suppose that that property exists and that *that it is a cube* does not exist. (And, in any case, where did *it*—that "second-order" property—come from? It seems that the defender of "property creationism" faces a vicious infinite regress.)

It is, moreover, hard to see what it could mean to speak of creating a "thing that can be said of things"—as opposed to creating things of which a certain thing that can be said of things is true. "Creation", after all, is a causal relation, and things that can be said *of* things (properties) can no more enter into causal relations than can things that can be said full stop (propositions). I conclude that properties—like propositions—are not created things.

3. Lightweight Platonism and Divine Aseity

In this section, I will examine an argument of Craig's for the conclusion that the existence of uncreated properties is inconsistent with the doctrine of divine **aseity**. I will contend that, if uncreated properties are "properties" in the sense in which I use the word, then one of the premises of Craig's argument is false.

This is Craig's argument:

> On Platonism, *deity* is an abstract object existing independently of God, to which God stands in the relation of exemplification or instantiation. Moreover, it is in virtue of standing in relation to this object that God is divine. He is God because He exemplifies *deity*. Thus, on Platonism, God does not really exist *a se* at all.
>
> (p. 43)

Well, no doubt that argument shows that *some* forms or versions of Platonism imply that God does not exist *a se*. Perhaps, indeed, true Platonism, the Platonism of Plato, implies that God (if he exists at all) does not exist *a se*. But it does not show that that is an implication of my lightweight platonism—which, remember, I spell with a lower-case 'p' precisely to distinguish it from Plato's theory of Forms.

Craig's argument relies on the assumption that Platonists are committed to a principle that lightweight platonists reject. This principle could be stated this way: Substitute any adjective you like for '*F*' in the following expression

> If an object is *F*, it is *F* because it has the property of being *F*,

and the result will be a truth.

(The word 'because' is highly ambiguous. It should be so understood that the offset expression means 'What it is for an object to be *F* is for it to have the property of being *F*.)

Let us call this principle **the Explanatory Principle**. The Explanatory Principle endorses propositions like these.

> If an object is white, it is white because it has whiteness (i.e., has the property of being white)

> If an object is wise, it is wise because it has wisdom (i.e., has the property of being wise)

> If an object is fragile, it is fragile because it has fragility (i.e., has the property of being fragile).

Lightweight platonists reject the Explanatory Principle.[6] They contend, in fact, that things are precisely the other way round: white things have whiteness because they are white; wise things have wisdom because they are wise; fragile things have fragility in virtue of their being fragile. For remember, for the lightweight platonist, a property is something that can be said of things—*that it is white*, for example. And for an object to have whiteness is for *that it is white* to be true of it. Is it not obvious that

> The Taj Mahal is white because *that it is white* is true of it

is false, and

6. A note for those who are familiar with Russell's Paradox. In any case, the quantifiers in the principle would have to be restricted in some way to avoid "Russellian" counterexamples. (Let the adjective 'non-self-applicable' be true of just those things that are properties that do not have themselves. The "unrestricted" principle and the true statement 'Fragility is non-self-applicable' together imply 'Fragility is non-self-applicable because it has non-self-applicability'. And the existence of non-self-applicability entails a contradiction.)

> *That it is white* is true of the Taj Mahal because the Taj Mahal is white

is true? Aristotle has said:

> It is not because we think truly that you are white [= pale], that you are white, but it is because you are white that we who say this have the truth.
> (*Metaph.* Theta/IX, 10)

Aristotle's thesis is very close to this thesis: It is not because *that Critias is white* is true that Critias is white; it is rather because Critias is white that *that Critias is white* is true. That thesis is a thesis about propositions. The corresponding or analogous thesis about properties is:

> It is not because *that it is white* is true of Critias that Critias is white; it is rather because Critias is white that *that it is white* is true of him.

In my view, 'Because *that it is white* is true of Critias' makes no more sense as an answer to the question, 'Why is Critias white?' than does 'Because *that Critias is white* is true'.

A re-writing of the above quotation from Craig so that it applied to lightweight platonism would be something like this.

> According to lightweight platonism, *that it is divine* is an abstract object existing independently of God. And *that it is divine* stands in the relation "is true of" to God. Moreover, it is because *that it is divine* stands in this relation to God that God is divine. He is God because *that it is divine* is true of him. Thus, according to lightweight platonism, God does not really exist *a se* at all.

And, should anyone ever actually say that the lightweight-platonist response would be simply to deny that the thesis 'God is God because *that it is divine* is true of him' is a consequence of the joint affirmation of theism and lightweight platonism. The lightweight platonist who is a theist rather affirms:

> *That it is divine* is true of God because God is God (that is, because God is divine)

—a statement that is strictly parallel to the following statement, and which no more has theological consequences than the following statement has anthropological consequences:

> *That it is human* is true of Socrates because Socrates is human.

I am, therefore, perfectly comfortable with saying both that things that can be said of things are uncreated *and* "We believe in one God, the Father, the Almighty, maker of heaven and earth, of all that is, seen and unseen" (and, a moment later, the words, "Through him all things were made"). I am comfortable affirming both that things that can be said of things are not brought into being by God *and* my allegiance to these words from the Prologue to the Fourth Gospel: "All things came into being through him and without him not one thing came into being."

An important element of the *credo* that I, as a Christian philosopher, bring to my vocation is this:

> If there were made known to me a compelling reason to believe that any of the propositions that I have defended in my philosophical writings contradicted the faith of the one, holy, Catholic and apostolic Church (and a proof that they contradicted the teachings of the Fathers would certainly be a significant step in the direction a compelling reason to believe it), I would immediately cease to defend (and cease to accept) those propositions. I would, in fact, publicly repudiate them.

I insist, however, that I have not yet seen any reason to believe that the following theses

> There exist both propositions (things that can be said, full stop) and properties (things that can be said *of* things). All propositions and properties exist necessarily, and, owing to the fact that they cannot enter into causal relations, are uncreated.

contradict the teachings of the Fathers. I have studied carefully the quotations from the Fathers that Craig has provided, and I can find nothing in them that is inconsistent with these theses. Everything said in those quotations seems to be addressed to issues entirely unrelated to them. I have given one example of this in the previous section (the quotation from Athenagoras of Athens). I will give a second, a single telling sentence taken from a much longer excerpt by Craig (p. 37) from a third-century dialogue called *On Free Will* (it has been ascribed to various authors). One of the participants speaks:

> Do you say then, that there co-exists with God matter without qualities out of which He formed the beginning of the world?

If we are interested only in the question whether the speaker is using the Greek word that has been translated 'qualities' to refer to items that in any way or respect resemble the items I call 'properties', we need read no further. It suffices to note that the author of *On Free Will* takes the question he has put into the speaker's mouth to be intelligible. And it would not be an intelligible question if "qualities" were things that could be said of things. For if qualities are things that can be said of things, and if matter is "without qualities," then matter is something of which nothing can be said. And if one says, "Matter is something of which nothing can be said," one has said something about matter. (One who says "Matter is something of which nothing can be said," can be compared to someone who says, "No one ever makes any statements" or "I am not here.") Therefore, whatever the question, 'Does there co-exist with God matter without qualities out of which. . . ?" means, it does not mean 'Does there co-exist with God something of which nothing can be said truly, namely matter, out of which . . . ?"

3.1 CODA

The arguments I have presented in this section and the previous section are (quite literally) arguments about nothing at all if there are no abstract objects. In Craig's view, my reasons for believing that there are abstract objects essentially incorporate a wrong **meta-ontology**—that is, wrong views on such matters as the nature of **quantification** and "ontological commitment." It is to this charge that Sections 3–8 of this essay are devoted.

Meta-ontology is the study or discipline that enquires into the nature of ontology (in any of the senses of 'ontology' outlined in the box on p. 19). *A* meta-ontology is a position in meta-ontology the discipline.

In the summary of Section 1 (p. 11) it was said that "Van Inwagen . . . suggests that the entities *he* calls abstract objects were unknown to the philosophers of the ancient world." In Section 2, some content is given to this suggestion. Perhaps the most important class of abstract objects—properties—are *things that can be said of things*. (One can, for example, say of Socrates that he is wise, and one can say the very same thing of Solomon.) The position is defended that these objects could not have figured in the thoughts of any philosopher or theologian of the ancient classical world. In Section 3, it is shown that Craig's argument for the conclusion that Platonism implies that God is not self-existent does not apply to a "lightweight" platonism—that is, to a platonism that identifies "properties" with things that can be said of things.

4. The Existential Quantifier and Ontological Commitment

4.1 Preliminary Remarks

Professor Craig and I disagree about a great many things in philosophy, but perhaps our most fundamental philosophical disagreement has to do with the relation between being (or existence) and the existential quantifier.

Quantifiers, Variables, and Ontological Commitment

Philosophers have found it useful to represent 'all' and 'some' statements—statements like 'Some Greeks are philosophers' and 'All Greeks are mortal' by means of the symbols '∃' (the **existential quantifier**) and '∀' (the **universal quantifier**). Thus, and '∃*x* (*x* is

Greek and *x* is a philosopher)' represents 'Some Greeks are philosophers', and '∀*x* (if *x* is Greek, then *x* is mortal)' represents 'All Greeks are mortal'. The letters '*x*' and '*y*' (and '*z*' and various other lowercase italic letters from the far end of the alphabet) are called **variables**. Variables are in a sense interchangeable—for we might as well have represented 'Some Greeks are philosophers' by '∃*y* (*y* is Greek and *y* is a philosopher)' or '∃z (z is Greek and z is a philosopher)'. There is more than one variable because more than one variable is sometimes needed inside a single sentence; for example, 'All Greeks have mothers' could be represented' as '∀*x* (if *x* is Greek, then ∃*y* (*y* is a mother of *x*))' or as '∀*y* (if *y* is Greek, then ∃z (z is a mother of *y*)).' Sentences containing the two quantifiers are pronounced in various ways when they are read aloud. For example, the second sentence above is sometimes pronounced 'For some z, z is a Greek and z is a philosopher', but it is also pronounced, 'There is a z such that z is a Greek and z is a philosopher' and 'There exists a z such that z is a Greek and z is a philosopher'.

If 'Some Greeks are philosophers' and 'There exist Greek philosophers' can both be represented by '∃*x* (*x* is Greek and *x* is a philosopher)', that fact strongly suggests that the former two sentences mean more or less the same thing. And if that is true, it suggests that 'Some of the things Kant believed are false' and 'There exist false things that Kant believed' mean more or less the same thing. And that, in its turn, suggests that 'Some of the things Kant believed are false' entails that there exist things that people believe—and what are things that people believe if not propositions? A person's **ontology** (from Greek, *on*, being) comprises those things whose existence he or she affirms. If a person must include Xs in his or her ontology if his or her total set of beliefs is to be consistent, we say that belief in Xs is among that person's **ontological commitments**. Is there a plausible case to be made for the thesis that someone who says 'Some of the things Kant believed are false' is ontologically committed to propositions? That will be our topic in the remainder of this chapter.

Although few philosophers would dispute much of what was said in the box, they do not agree about the meaning of sentences like '∃ *x* (*x* is Greek and *x* is a philosopher)'. There are various accounts of the meaning of '∃' in the philosophical literature. All these accounts are consistent with what was said in the box, and all of them respect the rules that govern logical inferences like

$$\exists x \forall y\ (x \text{ loves } y), \text{ therefore } \forall x \exists y\ (y \text{ loves } x)$$

(Something loves everything, therefore, everything is loved by something.) It would therefore be idle to dispute about which of these accounts is right. What I will do is to give my own account of the meaning of '∃', an account on which my position on the relation between being and '∃' will be based.

4.2 The Meaning of the Existential Quantifier

I will introduce an imaginary modification of English I will call **Revised English**. We begin by modifying the meaning of the pronoun 'it': we stipulate that in Revised English 'it' has no implications as to sex, gender, or personhood. Thus in revised English one could say 'If we hire a valet parking attendant, it will have to be able to drive a stick shift' and 'Every mother loves its children'.

Generally speaking, an occurrence of a third-person pronoun ('she', 'it', 'they', . . .) in a sentence has an **antecedent**—an earlier word or phrase that it "refers back to."

In Revised English,

—if an occurrence of 'it' in a sentence has an antecedent, its antecedent is in that same sentence

—an occurrence of 'it' has at most one antecedent (that is, it has either one or none)

> **Pronouns and Their Antecedents**
>
> In the following sentences, each pronoun is in bold-face, and its antecedent is underlined.
>
> Alice found Tom's wallet and returned **it**.
> Alice found Tom's wallet and phoned **him**.
> My dog saw a cat and began to chase **it**.
> Ask everyone whether **they** got **their** flu shot this year.
> There was something Jack didn't mention, and **it** was important.

—every antecedent of any occurrence of 'it' is an occurrence of the word 'something' or an occurrence of the word 'everything' (but in what immediately follows, we are going to ignore 'everything').

—the word 'something' occurs only in the phrase 'something is such that'. We call this phrase **the specific quantifier.**[7] If an occurrence of 'something' is the antecedent of an occurrence of 'it', we say that the occurrence of the specific quantifier of which that occurrence of 'something' is a part is the *quantifier-antecedent* of that occurrence of 'it'.

—every occurrence of the specific quantifier is followed by a sentence enclosed in round brackets. (Thus:

> Something is such that (it is a cat)
>
> Something is such that (it is a dog and it chases cats) and dogs are mammals.

The sentence surrounded by the round brackets following an occurrence of the specific quantifier is its **scope**. (We shall sometimes informally omit the brackets of there is only one way brackets could be re-introduced into the resulting sentence.)

—if an occurrence of the specific quantifier is the quantifier-antecedent of an occurrence of 'it', that occurrence of 'it' must fall within the scope of that occurrence of the specific quantifier. (So, for example, in the sentence

> Something is such that (it is a dog) and it chases cats.

the occurrence of the specific quantifier may be the quantifier-antecedent of the first of the two occurrences of 'it'—for that occurrence of 'it' falls within its scope. It cannot be the antecedent of the second occurrence, which does not fall within its scope.)

Now, except for the brackets Revised English is English, and we know what its sentences mean. If we wished, we could express sentences like

7. Strictly speaking, I need to say something about tense at this point. I need to say something about expressions like 'something was such that' and 'something will be such that'. But such a discussion would take us far out of the way, and I will leave tense in an intuitive and undefined condition in what follows.

My dog saw a cat and began to chase it

There was something Jack didn't mention, and it was important

using 'something is such that' and pronouns used according to the rules of Revised English:

Something is such that (it is a cat and my dog saw it and began to chase it)

Something is such that (Jack didn't mention it and it was important).

(As these two sentences show, it may be that more than one occurrence of 'it' falls within the scope of an occurrence of 'something is such that'.)

What we have said so far contains nothing that rules out sentences like

Something is such that (something is such that (it is a dog and it is a cat and it chased it till it was exhausted)).

And indeed we do not want to rule them out, for they are indispensable in science and everyday life. But in this sentence there are two occurrences of the specific quantifier and four occurrences of 'it'. And all four occurrence of 'it' fall within the scopes of *both* occurrences of the quantifier. How shall we determine which occurrences of 'something' are the antecedents of which occurrences of 'it'?

Essentially the same question can be asked about sentences of ordinary English. Consider, for example,

A dog will sometimes chase a cat till it is exhausted.

This sentence is ambiguous, for the antecedent of 'it' could be 'a dog' or it could be 'a cat'. One resource employed by ordinary English to resolve such ambiguity is this familiar device:

A dog will sometimes chase a cat till it (the dog) is exhausted

A dog will sometimes chase a cat till it (the cat) is exhausted.

But this device is awkward and can be applied only in a certain class of cases. One possible device for indicating the antecedents of pronouns unambiguously in Revised English is to make use of different fonts and sizes:

> **Something** is such that (*something* is such that (**it** is a dog and *it* is a cat and **it** chased *it* till *it* was exhausted)).

The idea, of course, is that an occurrence of 'something' is an antecedent of an occurrence of 'it' (that falls within the scope of the occurrence of the specific quantifier of which it is a part) if and only if they are written in the same font and size. We may contrast the offset sentence immediately above with

> *Something* is such that (**something** is such that (*it* is a dog and **it** is a cat and *it* chased **it** till *it* was exhausted)).

The former says that that a dog chased a cat till it (the cat) was exhausted. The latter says that that a dog chased a cat till it (the dog) was exhausted.

The main problem with using fonts and sizes of text in this way is that it is very difficult to apply in hand-written documents. A more practical device is this: place a little tag on occurrences of 'something' and occurrences of 'it'—and then, an occurrence of 'something' is an antecedent of an occurrence of 'it' (that falls within the scope of the occurrence of the specific quantifier of which it is a part) if and only if they bear the same tag. I will use the last few letters of the alphabet, written as superscripts, as tags. Thus:

> Something^{x} is such that (something^{y} is such that (it^{x} is a dog and it^{y} is a cat and it^{x} chased it^{y} till it^{y} was exhausted))

> Something^{x} is such that (something^{y} is such that (it^{x} is a dog and it^{y} is a cat and it^{x} chased it^{y} till it^{x} was exhausted)).

Some readers, at least, will have seen where we are going with these innovations. Our final step consists in introducing, by two abbreviations, *variables* ('x', 'y', 'z', . . .) and the *existential quantifier* ('$\exists$'). The first abbreviation is

> Replace 'it^{x}' with 'x'; replace 'it^{y}' with 'y' . . . and so on.

And the second is

> Replace 'something^x is such that' with '∃*x*'; replace 'something^y is such that' with '∃*y*' . . . and so on.

Applying these two abbreviations to the above two offset sentences, we have

> ∃*x* (∃*y* (*x* is a dog and *y* is a cat and *x* chased *y* till *y* was exhausted))

> ∃*x* (∃*y* (*x* is a dog and *y* is a cat and *x* chased *y* till *x* was exhausted)).

The universal quantifier (less important for our purposes) can be explained in essentially the same way: if the existential quantifier is an abbreviation of 'something is such that', the universal quantifier is an abbreviation of 'everything is such that'.

The process of abbreviation is easily reversed. Here is a sentence taken from the box **Quantifiers, variables, and ontological commitment** above:

> ∀*y* (if *y* is Greek, then ∃*z* (*z* is a mother of *y*)).

Those who ask, "What does that sentence mean?" are easily answered. To answer them, first substitute 'everything^y is such that' for '∀*y*' and 'something^z is such that' for '∃*z*' and 'it^y' for '*y*' and 'it^z' for '*z*'. The result of these substitutions is

> Everything^y is such that (if it^y is Greek, then something^z is such that (it^z is a mother of it^y)).

And one may say that that sentence expresses what '∀*y* (if *y* is Greek, then ∃*z* (*z* is a mother of *y*))' means—for the latter expression is nothing more than a convenient abbreviation of that sentence.

I claim in this section to have given an account of what the existential and universal quantifiers and the variables they "bind" mean. If someone asks why anyone should want there to be expressions that have the meanings I say quantifies and variables have, the answer (in a word) is "logic." Logic is the science of valid inference or valid argument. Consider for example this argument.

1. If two animals are conspecific, they have some of the same anatomical characteristics.

2. All dachshunds are animals and all Saint Bernards are animals.

3. Any dachshund and any Saint Bernard are conspecific.

4. There are dachshunds.

hence,

5. Some dachshund has some of the same anatomical characteristics as any Saint Bernard.

Is this argument **logically valid**? That is, does its **conclusion** (statement (5)) follow logically from its **premises** (statements (1) through (4))?

Arguments and Logical Validity

An **argument** is a sequence of two or more statements. The final statement of an argument is its **conclusion**, and the others its **premises**. If the conclusion of an argument can be deduced by correct reasoning from its premises, the argument is **logically valid**. It is important to note that an argument's being logically valid does not imply that its premises are true. Thus, both the following arguments are logically valid:

All Greeks are mortal There are Greeks	All Greeks are immortal There are Greeks
hence,	*hence,*
Some Greeks are mortal	Some Greeks are immortal.

(When writing an argument down so that it may be read and considered by others, it is customary, but not essential, to separate its conclusion from its premises by placing some word or phrase or symbol—such as '*hence*' or '*and therefore*' or '∴'—between the final premise and the conclusion.) If all the premises of a logically valid argument are true, that argument is **sound**. Thus, the argument on the left above is sound, and the argument on the right (although valid) is not sound.

The logic based on quantifiers and variables answers this question—or at least answers it given a "translation" of statements (1) through (5) into the quantifier-variable idiom. Here is one such translation—and it seems to be a perfectly reasonable one:

1′. ∀*x*∀*y* (if *x* is an animal and *y* is an animal and *x* and *y* are conspecific, then ∃*z* (*z* is an anatomical characteristic of *x* and *z* is an anatomical characteristic of *y*))

2′. ∀*x* (if *x* is a dachshund, *x* is an animal) and ∀*x* if *x* is a Saint Bernard, *x* is an animal

3′. ∀*x*∀*y* if *x* is a dachshund and *y* is a Saint Bernard, then *x* and *y* are conspecific

4′. ∃*x* *x* is a dachshund

hence,

5′. ∃*x* (*x* is a dachshund and ∀*y* (if *y* is a Saint Bernard, then ∃*z* (*z* is an anatomical characteristic of *x* and *z* is an anatomical characteristic of *y*)).

The techniques taught in any course on what is sometimes called "symbolic logic" show that (5′) can be validly deduced from (1′) to (4′) by a series of steps that could be checked by computer. (What these steps are is not a question we need consider.)

But those same techniques show that the statement

6. ∃*x*∃*y*∃*z* (*x* is an anatomical characteristic of *y* and *x* is an anatomical characteristic of *z*)

can also be deduced from (1′) and (2′) and (4′) and the presumably uncontroversial statement, '∃*x* *x* is a Saint Bernard'. The following three statements, moreover, are very easy to believe.

7. ∀*x*∀*y*∀*z* (if *x* is an anatomical characteristic of *y* and *x* is an anatomical characteristic of *z*, then *x* is an anatomical characteristic).

8. $\forall x$ if x is an anatomical characteristic, then x is a characteristic.

9. $\forall x$ if x is a characteristic, then x is a property.

As to statement (9), are not 'property', 'characteristic', 'feature', 'quality', and 'attribute' all more or less synonyms? Consider

> One of the most important properties of water is its ability to dissolve ionic and polar chemicals to form aqueous solutions.
>
> One of the most important characteristics of water is its ability to dissolve ionic and polar chemicals to form aqueous solutions.
>
> One of the most important features of water is its ability to dissolve ionic and polar chemicals to form aqueous solutions.
>
> One of the most important qualities of water is its ability to dissolve ionic and polar chemicals to form aqueous solutions.
>
> One of the most important attributes of water is its ability to dissolve ionic and polar chemicals to form aqueous solutions.

Note now that

10. $\exists x$ x is a property

follows from (6) to (9). And, if we like, we can add one further plausible premise to our argument:

11. $\forall x$ if x is a property, then x is an abstract object,

allowing us to deduce, finally,

12. $\exists x$ x is an abstract object.

And how shall we interpret sentence (11)? Our account of the existential quantifier tells us that (11) is an abbreviation of the Revised English sentence.

13. Something^{x} is such that (it^{x} is an abstract object).

But the devices peculiar to Revised English on exhibit in (13)—the "tags" and the brackets—are not really doing anything, for their purpose is to remove ambiguities of antecedence, and there are none in the English sentence

14. Something is such that it is an abstract object.

Sentence (11) means what sentence (12) means, and sentence (12) means what sentence (13) means. And all three mean the same as 'something is an abstract object'.

And now I will add a further thesis about meaning to those already affirmed in the present section, namely that all the sentences in the following list mean more or less the same thing:

$\exists x$ x is an abstract object

$\sim \forall x \sim x$ is an abstract object

Something is an abstract object

There is an abstract object

There are abstract objects

There are such things as abstract objects

It is not the case that everything is not an abstract object

The number of abstract objects is not zero

Abstract objects exist.

I said that my position on the relation between being and '∃' would be based on my account of the meaning of the existential quantifier. And that relation is explained by that "further thesis." For being (or existence) is what is expressed by sentences like those in the above list—although, of course, sentences "like" those in the above list need not concern the being of any particular being or the members of any particular category of being; the following sentences are also of this kind:

God exists

There are stable transuranic elements

There are 192 solutions to Heun's Equation.

Suppose my account of the existential quantifier is correct. Suppose all the sentences in the above list mean "more or less the same." (That is, suppose that they come at least as close to meaning the same thing as do, say, the two sentences 'The number of planets is not 0' and "The number of planets is greater than or equal to 1.") We may then conclude that (for example) 'God exists' differs in no important respect from '∃*x x* is God' and 'Something is God'.

Suppose someone were to say to me, "Well, I don't dispute the proposition '∃*x x* is the property "wisdom"'. But that's not a metaphysical proposition. What I (as metaphysician) want to know is the truth-value of a certain metaphysical proposition, a proposition that I might express in either of the two following ways:

> The property 'wisdom' exists in the strict and philosophical sense of 'exist'
>
> The property "wisdom" *is* in the metaphysically serious sense of 'is'."

My reply would be a simple one: 'exist' and (the existential) 'is' have only one sense, so there's no distinction to be made between the 'strict and philosophical sense' or the 'metaphysically serious sense' of those words and some other sense or senses. they might be supposed to have.

Let this much serve as a preliminary statement of my views on the existential quantifier and ontological commitment. I now turn to Craig's reasons for thinking that these views are mistaken. In the course of discussing and replying to Craig's arguments, I will say more about my own views as well.

The topics of Section 4 are the relation between being and the existential quantifier and the role the existential quantifier plays in determining a person's "ontological commitments." A large part of the section is devoted to an exposition and defense of an account of the meaning of '∃', the existential quantifier—that is, of the meaning of the sentences in which that symbol occurs, sentences like '∃*x x* is an abstract object'. It is maintained that that sentence means 'Something is such that it is an abstract object'. It is further maintained that those two sentences mean more or less the same as 'Not everything is not an abstract object', 'There are abstract objects', and 'Abstract objects exist'. It is contended that, therefore, a person is committed to the existence of abstract objects if '$\exists x$ x is an abstract object' is logically deducible from that person's beliefs—that is, that in that case the person is "ontologically committed" to abstract objects.

5. Craig on the Existential Quantifier and Ontological Commitment

Craig recommends that we "simply reject the view that existential quantifiers and singular terms are devices of ontological commitment" (p. 103). I will make only a passing remark about *singular terms* as "devices of ontological commitment." In my view—and here I follow W. V. Quine who coined the term 'ontological commitment'—an occurrence of a singular term in a sentence is relevant to "the ontological question" ('What is there?') only if it "occupies a position subject to existential generalization." To see what that means, consider the sentence

> The shadow of the flagpole was twice as long at 4:00 p.m. as it was at noon.

Does the fact that the term 'the shadow of the flagpole' occurs in the way it does in this sentence imply anything about what the sentence "says there is"? Only, I would say, if that term occupies a position subject to existential generalization. That is, only if

> $\exists x$ x was twice as long at 4:00 p.m. as x was at noon

(or 'Something is such that it was twice as long at 4:00 p.m. as it was at noon') is a logical consequence of that sentence. (In my view, it is not.) Ontological commitment, I maintain, is the business of quantifiers, and not of singular terms. Or, at any rate, it is only in a derived or vicarious sense that the singular terms that occur in a sentence are of any ontological consequence, a sense to be explained by reference to such existentially quantified sentences as may follow from the sentences in which they occur by existential generalization on their occurrences. Here endeth the passing remark concerning singular terms as "devices of ontological commitment."

I turn to the *interesting* part of Craig's proposal: that we reject the view that *existential quantifiers* are devices of ontological commitment. Interesting, to be sure, but I find it hard to understand, for it seems to me to rest on the assumption that we have some sort of "pre-analytic" grasp of the concept of ontological commitment, a grasp that we have independently of and prior to our understanding of the existential quantifier. But the phrase 'ontological commitment' is not a phrase of ordinary language (like such philosophically important words as 'know' and 'cause' and 'possible'); it is rather a *pure term of art*, and therefore means whatever its inventor (Professor Quine) says it means. And the existential quantifier was a central and essential component of Quine's explanation of the meaning of his term of art.

What, then, does the person who says, "The existential quantifier is a not a device of ontological commitment," saying? What does that statement *mean*? Quine, as far as I know, did not use the phrase 'device of ontological commitment', and Craig does not define the term.

Now it happens that I myself am willing to say that the existential quantifier *is* a device of ontological commitment—but only if that statement is taken in very carefully specified sense. The purpose of the present section is to specify (carefully) that sense.

In Section 4.2, I said that the phrase 'something is such that' was "the specific quantifier" and that the symbol '∃' was the "existential quantifier." In the present section I am going to modify that usage and speak of 'something is such that' as the existential quantifier. I do this because it is really that phrase (or any equivalent English phrase—or any equivalent phrase of some other language, such as '*quelque chose est tel que*' or '*etwas ist derart, daß*') that is a "device of ontological commitment." The symbol '∃' is, as I see matters, simply an abbreviation of 'something is such that' or some equivalent English phrase (or phrase of French or German or Xhosa or Iroquois . . .). And, therefore, it is a device of ontological commitment if and only if 'something is such that' is a device of ontological commitment. (The same point of course applies to the other symbols that various writers on logic have used with the same meaning as '∃', such as 'V' and 'Σ'.) The important thing is not the arbitrarily chosen symbol that might be used to abbreviate 'something is such that'. but the phrase itself: if I may so express myself, the *real* existential quantifier is 'something is such that' (or some more or less equivalent phrase, such as 'it is true of at least one thing that').

I will offer a definition of the sentential schema,

S *A* is ontologically committed to ___s.

(The definition will, I believe, display the way in which the existential quantifier—'something is such that'—figures in the concept of ontological commitment. It will display the sense in which 'something is such that' is a "device of ontological commitment.")

Instances of the schema S are obtained by replacing '*A*' with a word or phrase that designates a person and replacing '___s' with a plural **count-noun** or nominal phrase.

Count-Noun

A count-noun is a noun that can be modified by 'a(n)', the indefinite article, and has a plural form. Thus 'dog' is a count-noun, for we can say, '*A* dog is sometimes hostile to all other dog*s*'. Many nouns are count-nouns in some contexts and not in others. For example, 'salt' is a count-noun in the statement 'A salt is a compound of a positive and a negative ion', and not a count-noun in the statement 'His doctor told him to use less salt'; 'democracy' is a count-noun in the statement 'A democracy is a fragile thing', and not a count-noun in the statement 'Democracy was invented by the ancient Greeks'.

Thus, 'Plato is ontologically committed to Forms' and 'Aristotle is ontologically committed to substances' (and, for that matter, 'Vladimir Putin is ontologically committed to hobbits') are instances of S.

I stipulate, however, that we shall be interested in questions about a person's ontological commitment "to ___s" only in cases in which, to speak loosely, we really know what we're talking about when we talk about ___s. Suppose, for example, you ask me whether I'm ontologically committed to substances, and I reply, "Well, that all depends on what you take 'substances' to be—on what you think the nature of a substance is, on what you suppose substances to be *like*." I proceed to inform you that some "substance theorists" say that substances are things with the properties F, G, and H. Others, however, contradict those theorists and staunchly maintain that if there are substances, they indeed have the properties F and G, but lack the property H. I further inform you that the disagreement between the two schools of substance theory is a metaphysical disagreement—that is, it cannot be resolved either by the methods of the empirical sciences or by linguistic analysis, for both theories about the nature of substances are consistent both with all possible observations and experimental results and with the meaning of the English word 'substance'. I insist, therefore, that when you ask me whether I'm ontologically committed to substances, the only reply that I—who am of the "F and G but not H" school—can safely make is that my ontology includes items that have both F and G, but nothing that has F, G, *and* H.

The point of my stipulation is trivial, really: If you and I are debating about whether Miriam is ontologically committed to the existence of substances (in the sense in which *you and I* use the word 'substance'—and you and I had better be using it in the *same* sense or we shall be talking past each other—and not *Miriam's* sense of 'substance' if that sense should be different from ours), our debate makes sense only to the extent that the two of us agree not only about what the word 'substance' *means*, but also agree about what substances are or would be if there were such entities—about what properties substances have or are supposed to have.

All right—back to the task of finding a definiens for S. I begin with a definition of "*direct* ontological commitment":

> *A* is directly ontologically committed to ___s $=_{df}$ *A* believes that something is such that it is a(n) ___.

('*A*' and '___s' are to be understood as in the previous schema; '___' is to be replaced by the singular count-noun or nominal phrase whose plural has replaced '___s' in the **definiendum**.)

Definition, definiendum, definiens

We illustrate the correct application of these terms to

> *x* is a *triangle* $=_{df}$ *x* is a polygon with three edges and three vertices

The whole expression is a **definition**. '*x* is a *triangle' is the* ***definiendum*** *(Latin: the "to be defined") of this definition, and* '*x* is a polygon with three edges and three vertices' is its **definiens** (Latin: the "defining"). The plural of 'definiendum' is 'definienda'. The plural of 'definiens' is 'definientia'.

Thus, 'Aristotle is directly ontologically committed to substances' means 'Aristotle believes that something is such that it is a substance'.

The words 'is such that' occur in the definiens only because those words occur in our explanation of the quantifiers. In what follows I will generally omit them and write, e.g., 'Aristotle believes that something is a substance'.

Now some may find this definition puzzling. I am thinking of people who would say things like

> But consider statements of this form: "I believe that there are, strictly and literally speaking, ___s; that is, I believe that there are such *objects* or *entities* as ___s"—would not that be a better form of words to use to express ontological commitment ("direct" or otherwise) to substances than "I believe that something is a ___"?

If someone did make this protest, my reply would be,

> No, it would not. An equally good form of words, perhaps, but certainly not a better form of words. For I take, for example, 'There are, strictly and literally speaking, substances; that is, there are such *objects* or *entities* as substances' and 'Something is a substance' to express the same proposition. The principal difference between the two statements is that 'I believe that something is a substance' is a simple, straightforward statement and the longer, more elaborate statement is not. They are related in much the same way as 'Henry loves his mother' and 'Henry is bound to the maternal author of his being by adamantine chains of filial devotion'.

"But why does your definition of direct ontological commitment have such an elaborate definiens? Why not simply '*A* believes that there are ___s' or '*A* believes that ___s exist'?"

My answer to this question will involve reference to the following story. Let us suppose that Gerard Manley Hopkins has entered the following *pensée* in his journal:[8]

> If the world were emptied of light, it would be emptied of shadow as well—and I should mourn the loss of shadow almost

8. This example is borrowed from Van Inwagen (2014).

> as much as I should mourn the loss of light. But light exists! Shadows exist! All praise be to thee, O Lord, who hast created both light and shadow!

Consider the occurrence of the sentence 'Shadows exist' in this passage. Call the proposition expressed by 'Shadows exist' (there, in that particular journal-entry) the Hopkinsonian Proposition. I contend that the Hopkinsonian Proposition is *true*: and not merely "correct" or "nearly as good as true" or "true in the loose and popular sense but not in the strict and philosophical sense," but true *simplicter*, true *full stop*, true *period*. And all that is required for the truth of that rather undemanding proposition (I contend) is that various portions of the surfaces of things be shaded or in shadow. I accept the Hopkinsonian Proposition. And since there are contexts in which 'Shadows exist' expresses the Hopkinsonian Proposition, there are contexts in which it is true to say, "Van Inwagen believes that shadows exist." (After all, I believe that on occasion various portions of the surfaces of things are in shadow.)

But what about the sentence 'Something is such that it is a shadow'? In my view, if someone used that sentence to make an assertion, that person would therewith undertake an obligation to at least attempt to answer certain *metaphysical* questions, questions like,

> But what *are* shadows? What is their *nature*? What *properties* do they have? For example, can a shadow move across a surface? Can a shadow grow larger or smaller? Is the shadow a tree casts in sunlight identical with the shadow it casts by moonlight? Suppose we find the following statement in an astronomy textbook: 'During a solar eclipse visible at the equator, the moon's umbra or shadow is moving across the surface of the earth with a speed of 1270 kph'. Does this statement logically imply that during a solar eclipse visible at the equator, *something* is moving across the surface of the earth with a speed of 1270 kph? If the shadow of the flagpole was 20 meters long at noon and 40 meters long at 4:00 p.m., does it follow that *something* was twice as long at 4:00 p.m. as it was at noon?

Craig has said that ordinary speakers who say things of the forms '*F*s exist' and "There are *F*s" do not always suppose that they are "making ontological commitments" in any sense that would require them to answer pointed metaphysical questions concerning the properties of *F*s. I am in substantial agreement with that statement.

Suppose, however, that we define a conversational circumstance we call *being in the* **Ontology Room**. The conditions that define this conversational circumstance are simple: a conversation takes place in the Ontology Room, if and only if

(i) All parties to the conversation agree to say only what is strictly and literally true: in the Ontology Room we eschew all metaphors, manners of speaking, figures of speech, hyperbole . . . in the Ontology Room, all speakers are Russellian Pedants: people who like what they say to be true.

(ii) In the Ontology Room, whenever one makes a singular existential assertion in English, these assertions are to be regarded as abbreviations of the corresponding sentences of Revised English. So for example, 'God exists' is to be understood as an abbreviation of 'Something^x is such that (it^x is God)', 'There was such a person as Homer' is to be understood as an abbreviation of 'Something^x was such that (it^x was Homer)', and 'Robin Hood really existed' as an abbreviation of 'Something^x was such that (it^x was Robin Hood)'.

(iii) In the Ontology Room, whenever one makes existential assertions in English involving singular or plural count-nouns ('Greek', 'athletes', 'number', etc.) or count-noun phrases ('Athenian philosopher', 'Australian female professional basketball players', 'multiple of π') they are to be understood as abbreviations of phrases involving the existential quantifier. I provide illustrations of the way various existential statements involving the count-noun 'Greek' are to be understood:

A Greek exists	There exists a Greek	There is a Greek
Greeks exist	There exist Greeks	There are Greeks,

these assertions are to be regarded as having the following meaning:

Something^x is such that (it^x is a Greek).

(iv) And, finally, if one says (in English) something of the form

Some *F*s are *G*s

in the Ontology Room, one's statement is to be understood as an abbreviation of the corresponding statement of the form:

Something^{x} is such that (it^{x} is a(n) *F* and it^{x} is a(n) *G*).

(For example, if one says, 'Some Greeks are philosophers' in the Ontology Room, this is to be understood as an abbreviation of

Something^{x} is such that (it^{x} is a Greek and it^{x} is a philosopher).

(v) These are the only English existential idioms permitted in the ontology room. For example, one couldn't say this in the Ontology Room:

Last night I saw upon the stair.
A little man who wasn't there.
He wasn't there again today.
Oh, how I wish he'd go away...

One would have to replace the second and third lines of the verse with 'A little man who didn't exist. He didn't exist again today.' Or 'A little man and there was no little man upon the stair. There was no little man upon the stair again today' or some such. At any rate 'wasn't there' is not one of the existential idioms permitted in the Ontology Room.

This completes the statement of the conditions that define the conversational circumstance called 'being in the Ontology Room': A conversation takes place "in the Ontology Room" if and only if all the participants in that conversation agree to observe the requirements set out in the above statement. If I say that someone says something "in the Ontology Room," I mean that that person says that thing while a participant in a conversation that takes place in the Ontology Room.[9]

---

9. The debate between "Argle" and "Bargle" in David and Stephanie Lewis's "Holes" (Lewis and Lewis, 1983) is an excellent introduction to the conditions that define the Ontology Room. Every aspect of that debate conforms perfectly to those conditions.

I insist that if one does say any of the following things *in the Ontology Room*

The shadow of the flagpole exists	A shadow exists	Shadows exist
There exists a shadow	There exist shadows	
There is a shadow	There are shadows	

one is obliged to answer (or at least to attempt to answer) metaphysical questions about shadows—questions like the questions in the offset passage above. And this is because the rules that constitute the Ontology Room require the first of these sentences to be used with the meaning 'Something^x is such that (it^x is a shadow and it is cast by the flagpole)', and the others to be used with the meaning 'Something^x is such that (it^x is a shadow)'.

Outside the Ontology Room—in the ordinary business of life—no such rule is in force. (I don't suppose that any speaker engaged in the ordinary business of life would utter a sentence that was the result of appending a singular count-noun to 'Something is such that it is a'. If I did hear a sentence of that form uttered in a florist's shop or a law office or a lecture on the history of Chinese pottery, I'm not sure what I should conclude. (Possibly that the speaker had inexplicably veered from selling flowers or preparing a will or discussing pre-dynastic bowls and vases into metaphysics.) I maintain that outside the Ontology Room, utterances of 'Shadows exist' (or of 'A shadow exists' or 'There exists a shadow', etc.) express nothing more than the Hopkinsonian Proposition—that is, more or less, that opaque objects sometimes come between sources of light and some of the surfaces on which light from those sources would have fallen were it not for the interposition of those opaque objects. And the Hopkinsonian Proposition does not does entail the proposition that something^x is such that (it^x is a shadow). The former proposition, the Hopkinsonian Proposition, entails the existence of nothing but material objects (at least some of which are opaque) and light and sources of light. The latter proposition, however, is true only if *that it is a shadow* can be said truly of something, and *that it is a shadow* cannot be said truly of any material object or of light or of any source of light.

"Well, you're right about what 'most people' wouldn't say. Most people don't ever use words like 'Something is such that.'

(Most people don't speak English at all, of course. But I suppose you mean your thesis to apply to all speakers of languages that contain expressions equivalent in meaning to 'something is such that'—for example, '*quelque chose est tel que*' or '*etwas ist derart, daß*'.) So how do you know what your average hairdresser or bus driver is ontologically committed to? To take a particular example, how would you determine whether David the hairdresser believes that something is such that it is a shadow?"

That is a good question. My answer to this good question assumes the truth of the following thesis about belief:

> For any sentences *x* and *y*, if *x* and *y* express the same proposition, the predicates ⌜believes that *x*⌝ and ⌜believes that *y*⌝ are true of exactly the same things.[10]

Thus, for example, if 'All Cretans are liars' and 'All Cretans are mendacious' express the same proposition, then 'believes that all Cretans are liars' is true of Epimenides if and only if 'believes that all Cretans are mendacious' is true of him.

We are now prepared to discuss David. Suppose that, if you asked David whether he thought that something was such that it was a shadow, and that you got no response but "What?" accompanied by a blank look, it doesn't follow that David doesn't believe that something is such that it's a shadow. For it may be that there is some sentence that has this property: if David were to speak *that* sentence and I were to say "Something is such that it is a shadow," he and I would be saying the same thing in different words. Suppose that sentence were 'Shadows are real things'. In that case, it follows from our principle that if 'believes that shadows are real things' is true of David, then 'believes that something is such that it is a shadow' is also true of him. And it follows that if he says, "I believe that shadows are real things" (and believes what he says), he is directly ontologically committed to shadows.

Now let us consider any given person's direct ontological commitments—*all* of them—his or her direct ontological commitments to the existence of (as it might be) cats, shadows, attributes,

10. The reader who is unfamiliar with the odd-looking quotation marks ("Quine corners") in this principle may read the principle this way: 'If two sentences express the same proposition, the predicate formed by prefixing one of them with "believes that" and the predicate formed by prefixing the other with "believes that" are true of exactly the same things'.

divine beings, Cartesian mental substances, proper temporal parts of persisting material things, etc. Those commitments may be "collected" into a single proposition that I will call that person's direct ontological compendium—a proposition that could be expressed by a very long conjunctive sentence one small segment of which might be

> . . . something is such that it is a cat AND something is such that it is a shadow AND something is such that it is an attribute AND something is such that it is a divine being AND something is such that it is a Cartesian mental substance AND something is such that it is a proper temporal part of a persisting material thing . . .

And let us also consider the *non-existential* propositions that person accepts—propositions that do not logically imply the existence of anything.[11] That category will include universal propositions: that everything is such that it is partly equine if it is a unicorn, for example, or that everything is such that, if it is an attribute, it lacks causal powers, or that everything is such that it is not a unicorn. It will also include conditional propositions: that there are planets if there are moons, or that if there had been no mass extinctions, there would be no vertebrates. We may then—*finally*!—proceed to the following schematic definition:

> *A* is ontologically committed to ___s $=_{df}$
>
> That something is such that it is a(n) ___ is a formal, logical consequence of A's direct ontological compendium and the non-existential propositions that A accepts.

(I will sometimes use 'A's ontology includes ___s' as a stylistic variant on 'A is ontologically committed to ___s'.)

11. (Anyone who doesn't see what the point of this footnote is may safely ignore it.) Or, more exactly, propositions whose only existential implications are those they share with 'There is something'. (For I do not choose to burden myself with adherence to a "free" logic.) This will have the consequence that our definition of ontological commitment (a few lines down in the text) will entail that everyone is ontologically committed to self-identical things. (It will, moreover, entail that anyone who accepts 'Everything is such that it is partly equine if it is a unicorn' is ontologically committed to things that are partly equine if they are unicorns.) The difference between 'proposition that has no existential implications' and 'proposition that has no existential implications beyond those it shares with "There is something"' seems to me to be small enough that I am willing to ignore it in the text.

Here is a "toy" example of the application of this definition. Suppose that Willard lacks the belief that something is such that it is a lion. (He knows the word 'lion' but isn't sure whether lions—and indeed all supposed members of the order *Carnivora*—are real or mythical.) Then he is not directly ontologically committed to the existence of lions. But suppose he is directly ontologically committed to the existence of gazelles—that is, he believes that something is such that it is a gazelle. And suppose that he accepts the following non-existential propositions:

> If there are any carnivores at all, there are lions
>
> If there are herbivores, there are carnivores
>
> All gazelles are herbivores.

It follows logically from the four beliefs we have ascribed to Willard that there are lions, and he is therefore ontologically committed to lions—despite his not being *directly* ontologically committed to lions. (No doubt he *should* be directly ontologically committed to lions—no doubt he *should* believe that something is such that it is a lion—, for the existence of lions is an obvious logical consequence of his beliefs. But people do not always believe what they should believe, and they do not always accept the obvious logical consequences of their beliefs.)

This was, as I said, only a toy example of a case of ontological commitment that is not a case of direct ontological commitment. Realistic examples of such cases are harder to come by. I will give a realistic example, but, unfortunately, it will be understood only by readers who know a little set theory.

> Sally lacks the belief that something is such that it is a well-ordering relation on the real numbers, but accepts all the axioms that define the properties of the real line and accepts the axioms of some "first-order" set theory (both those that have and those that do not have existential implications) including the Axiom of Choice. Then she is ontologically committed to the existence of well-ordering relations on the real numbers—but is not directly ontologically committed to the existence of well-ordering relations on the real numbers.

My answer to the question 'Is the existential quantifier—that is, the phrase "something is such that"—a device of ontological commitment?' is this: 'Yes, but only in the sense provided by the following statements. A person's ontological commitments are determined by that person's direct ontological compendium (together with that person's non-existential beliefs); A person's direct ontological compendium consists of the propositions that person accepts that are expressed by sentences that are instances of the schema "Something is such that it is a(n) ___".'

It should be noted that there is a great deal of idealization in the notion of a person who has a direct ontological compendium. Few people actually use phrases like 'something is such that', and it is often a philosophical question which of several propositions a speaker has affirmed by the use of a sentence that contains the word 'exist' or phrases like 'there is a', and 'there exist'. For example, consider the occasion on which Gerard Manley Hopkins wrote the sentence 'Shadows exist!' in his journal. Does that sentence (in that context) express the proposition

> Various portions of the surfaces of things are shaded or in shadow?

Or does it express the proposition

> Something^x is such that (it^x is a shadow)?

These are hardly the same proposition. The latter is true if and only if there are things of which 'it is a shadow' can be truly said. And the former can be true if there are no things of which 'it is a shadow' can be truly said. (It is not clear what, exactly, the existential implications of the former are, but suppose that this is a correct list of them: physical objects, surfaces of physical objects and parts of the surfaces of physical objects, and light. It is clear that 'it is a shadow' is not true of physical objects or of light. Might it be true of parts of the surfaces of physical objects? Well, not if shadows can move across the surfaces of physical objects, and not if shadows can grow longer or shorter without any change in the physical objects on which they are cast.) I say it expresses only the former. But—this is my impression—some philosophers would say it expresses the latter. If so, it is a philosophical question whether the latter proposition belongs to Hopkins's direct ontological compendium.

Craig has denied that the existential quantifier is a "device of ontological commitment." In Section 5, it is contended that 'something is such that' is a device of ontological commitment in the following sense:

> Let us say that a person's *direct ontological compendium* contains (for example) abstract objects if that person is willing to say 'Something is such that it is an abstract object' and contains shadows if he or she is willing to say 'Something is such that it is a shadow' and so on. Then a person is *ontologically committed to* (for example) material objects if 'Something is such that it is a material object' is a logical consequence of that person's direct ontological compendium together with his or her non-existential beliefs.

6. Has 'There Is' "Lightweight" and "Heavyweight" Uses?

At the beginning of Section 4, I noted that Craig recommends that we "simply reject the view that existential quantifiers . . . are devices of ontological commitment" (p. 103). His recommendation is based on his conviction that the 'there is' of ordinary English can be used in a "lightweight" way (used in this way, it is not a device of ontological commitment) and in a "heavyweight" way (used in that way, it is a device of ontological commitment). And the existential quantifier—that is, the symbol '∃', not the phrase 'something is such that'—is neutral between these two ways of using 'there is'.

In this section, I will examine this idea.

I ask you first to consider two sentences from that monumental work *Systematic Theology*, which was written by my good friend and colleague Theodora the theologian:

1. There is but one living and true God, everlasting, without body, parts, or passions

2. There is but one path to righteousness, and that is faith in Jesus Christ.

Both sentences start with the words 'there is' (and indeed with the words 'there is but one'). It would obviously be right to say that in writing (1) Theodora affirmed the existence of a certain *object*[12]—an object that was everlasting, bodiless, mereologically simple, and impassible. And it would just as obviously be wrong, very wrong indeed, to say that in writing (2), she affirmed the existence of a certain object—an object that is a path to righteousness and is identical with faith in Jesus Christ. I accept without reservation the following statement:

> When Theodora wrote sentence (1), she affirmed the existence of an object—an object of which such things as *that it exists at all times* and *that it is not a composite, not a sum of parts* are true. But when she wrote sentence (2) she did not affirm the existence of an object. Or if she affirmed the existence of Jesus Christ, she at any rate did not affirm the existence of an object of which *that it is the only path to righteousness* is true (either in the literal sense of 'path' or in any metaphorical sense). The words 'there is' are therefore not always a "device of ontological commitment."[13] And the same point applies to the other informal quantifiers of ordinary language. It applies even to 'exists'—for Theodora could as well have written 'Only one path to righteousness exists.'

This statement perhaps raises the question, "How, then, if one encounters a sentence that contains 'there is' or 'there are', does one determine whether that sentence is an assertion of the existence of an object (or objects) of some specified kind?" (In this question, '"there is' or 'there are'" stands in for the whole—rather indeterminate—list of informal quantifiers of ordinary language that can, in various contexts, perform the same functions as 'there is'.) And my answer is, "Why, by using the methods that one normally employs to determine, in the conversational situation in which one finds oneself, the meaning of a statement that might be interpreted in more than one way—attention to context, common sense, and so on."

It is wholly evident, I think, that it would make sense for a reader of *Systematic Theology* to respond to Theodora's sentence (1) by asking questions like,

---

12. In the very abstract sense of 'object' employed by mathematicians and philosophers: 'object' is the most general count-noun.
13. This statement does not contradict the carefully stated thesis ('The existential quantifier is a device of ontological commitment') that was defended in the previous section.

> Theodora, would you be so kind as to explain to me how a thing can be "living" if it has no body and is not even a composite object?
>
> Tell me, Theodora, does the existence of an everlasting being imply that time had no beginning?

And it is equally evident that it makes *no* sense to ask questions like the following in response to Theodora's sentence (2):

> Theodora, I'm really interested in the ontology that underlies that thesis. Can you fill me in on what a "path" is? Obviously, it's not a garden-variety path on some surface, such as the surface of the earth. Are paths then abstract objects of some sort?
>
> So, Theodora, is it your position that there's a dyadic relation "leading to" and two objects, faith in Jesus Christ and righteousness such that the former bears that relation to the latter? And that nothing other than the former bears it to the latter?

It is, then, a fact about our discourse that sentences that contain words and phrases like 'exist' and 'there is a' are of two kinds: Theodora's sentence (1) is paradigmatic of one of these kinds, and her sentence (2) is paradigmatic of the other. I will call this fact the Two Kinds Fact. I hesitate, however, to give names to the two kinds, for fear that any of the names that tempt me might be regarded by philosophers who differ from me in the matter of how the Two Kinds Fact is to be explained as tendentious or question-begging.

The Two Kinds Fact calls for explanation, and more than one explanation is possible. Craig, following Jody Azzouni, Thomas Hofweber, and Robert M. Adams[14] explains the Two Kinds Fact by postulating that expressions like 'there is an ___' and '___ exists' and 'there are ___s' and '___s exist' have both a "heavyweight" or "existentially loaded" use and a "lightweight" or "ontologically neutral" use. In Theodora's sentence (1), for example, 'there is' is being used in the heavyweight way in and in her sentence (2) it is being used in the lightweight way. Only the heavyweight use of

14. Azzouni (2004); Hofweber (2016); Adams (2012).

'there is' "carries ontological commitment"; if one says, e.g., 'There are philosophical positions that I find it hard to take seriously', one is thereby committed to answering pointed metaphysical questions about the properties of "philosophical positions" only if in uttering that sentence one was using 'there is' in the heavyweight way—which one pretty obviously was not. Moreover, if one has said, "There are philosophical positions that I find it hard to take seriously" one must answer criticisms like, "You say you're a nominalist, but you've affirmed the existence of philosophical positions, which, whatever they may be, are certainly not concrete particulars" only if one has used 'there are' in its heavyweight sense.

Consider once more these two sentences

1. There is but one living and true God, everlasting, without body, parts, or passions

2. There is but one path to righteousness, and that is faith in Jesus Christ.

Why is it that it makes sense to ask someone who has uttered (1) how it can be that God is "living" if he has no body, and it does not make sense to ask someone who has uttered (2) what kind of entity a "path to righteousness" is? More generally, we have the Two Kinds Fact: "there is" statements are of two kinds. (Statement (1) is of one of these kinds, and statement (2) of the other.) But what is the difference between statements in the one category and statements in the other? According to the "two uses" party, the difference lies entirely within the phrase "there is." There are two ways to use this phrase: the way that is ontologically committing and the way that isn't. It's being used in the committing way in (1) and in the non-committing way in (2). The person who has uttered (1) has asserted the existence of God, and has thereby incurred a responsibility to answer various questions about the properties of God. The person who has uttered (2) has not asserted the existence of a path to righteousness, and so has not thereby incurred any responsibility to answer questions about the properties of a path to righteousness.

I would propose a different explanation of the Two Kinds Fact.[15]

15. I have been known to insist that explanations have no place in metaphysics. My thesis was that it is a mistake for metaphysicians to point to certain (alleged) facts and argue that a certain metaphysical theory should be accepted because it

There are two ways one might regard the syntactical structure of a sentence of ordinary English that contained "existential" words and phrases—'exist' and 'there is' and so on. First, one might take the sentence's syntactical structure as a serious guide to the way one should represent the proposition it expresses in the quantifier-variable idiom (or, what is the same thing, in Revised English). Secondly, well, one might not. Consider, once more, Theodora's two sentences. If her sentence (1) is rendered into Revised English, the translation will contain the existential quantifier ('Something is such that'), and the syntax of the translation can very nearly be "read off" the syntax of (1). And that translation would have to be something very much like this:

1′. Something^{x} is such that (it^{x} is divine and everything^{y} is such that (if it^{y} is divine, then it^{y} = it^{x})).[16]

(Here, 'is divine' abbreviates 'is a living, true, and everlasting God, who is without body, parts, or passions'.) But the sentence 'There is but one path to righteousness, and that is faith in Jesus Christ' cannot be translated into Revised English in the following way:

2′. Something^{x} is such that (it^{x} is a path and it^{x} leads to righteousness and everything^{y} is such that (if it^{y} is a path and it^{y} leads to righteousness, then it^{y} = it^{x}) and it^{x} = faith in Jesus Christ).[17]

If someone did speak or write (2′), it would be entirely in order to respond to that assertion by asking questions like, "But what are these 'paths'? Are they abstract objects of some sort?" If, moreover, one spoke the sentence (2) in the Ontology Room, it would be entirely in order to respond to that assertion by asking the same questions, for according to the conventions that constitute the Ontology Room, sentence (2), if spoken in the Ontology Room,

explains those facts better than all other theories that (supposedly) explain them. That is, I was deprecating the employment of "inference to the best explanation" by metaphysicians. I am not now repudiating my rejection of inference to the best explanation in metaphysics. I am not arguing that my position is to be accepted because it explains our data better than any of its rivals. I am simply pointing out that there *is* an explanation of those data that is consistent with my position.

16. That is to say: $\exists x$ (x is divine & $\forall y$ (y is divine $\rightarrow y = x$)).
17. That is to say: $\exists x$ (x is a path & x leads to righteousness & $\forall y$ (y is a path & y leads to righteousness. $\rightarrow y = x$) & x = faith in Jesus Christ).

must mean (2′). If one were a participant in a discussion in the Ontology Room and one wished to affirm the thesis that would be expressed by an utterance of sentence (2) outside the Ontology Room—a thesis that obviously does not commit those who affirm it to answering metaphysical questions concerning objects called 'paths'—one would have to abandon the words 'there is' and abandon the metaphorical use of 'path' and say something resembling.

2″. Everything^{x} is such that (if it^{x} is human and it is not the case that it^{x} is righteous, then it^{x} will become righteous if and only if it^{x} has faith in Jesus Christ).[18]

So—*au fond*—expressions like 'there is' and 'there exists' and 'some ___ is' and 'something is' cannot be said to have different uses or to be usable in different ways. But *sentences* that contain these expressions can be understood in ways that treat the role they play in those sentences differently. Revised English is a language that, by design, contains only one such expression, and—more importantly—forces that expression to conform to simple and precise syntactical rules. The difference between the role of 'there is' in sentence (1) and its role in (2) is simply that the Revised English translation of (1) has a syntax that more or less "maps" the English 'there is' on to the quantifier 'something is such that', whereas this is not the case with sentence (2): the syntax of any plausible Revised English translation of (2) will have little in common with the syntax of (2).

I would note, finally, that the opposition between taking the syntax of a sentence of natural language seriously as a guide to the way to translate it into an idiom whose syntax is more sharply defined (on the one hand) and not taking it as a serious guide (on the other) is not confined to questions concerning the ontological commitments one undertakes by using that sentence. For example: Carla the Cartesian says, "If I think, I necessarily exist." Literal Larry responds by saying, "But Carla, you do think. Surely you see that it follows by Modus Ponens from that fact and your statement

18. More cautiously: (2″) constitutes *one* proposal about how one might best express the proposition expressed by (2) in Revised English. It is, of course, often a vexed question whether a certain sentence of Language A is a satisfactory translation into A of a certain sentence of Language B.

that you necessarily exist? I thought only God could do that." Or Salt-of-the-Earth Sally says, "I never told a lie to nobody." And Larry proclaims, "Sally has confessed that there never was a time when she wasn't lying to somebody."

By way of summary:

I deny that the ordinary-language *operator* 'there is' has both a "heavyweight" (or "existentially loaded") use and a "lightweight" (or "ontologically non-committal") use. I affirm that many ordinary-language *sentences in which that operator occurs* can have both heavyweight and lightweight *readings*. (A heavyweight reading of an English sentence containing 'there is' is a reading that requires its translation into Revised English to have a syntax that corresponds closely to the syntax of the original. A lightweight reading is a reading that does not impose that requirement on its translation.) That is to say, one may take the syntax of an English sentence containing 'there is' *seriously*—so seriously as to grant it the authority to determine the syntax of the sentence one will use in its stead in the Ontology Room—or one may decline to take it seriously. And to "decline to take it seriously" is simply to decide *not* to grant it that authority.

And how does one determine, on any given occasion, whether an English sentence one has encountered—a sentence that contains an expression that is one of 'there is', 'exists' *et al.*—should receive a heavyweight or a lightweight reading? Well, how does one (if one is of the Azzouni-Hofweber-Adams-Craig school) decide whether an occurrence of 'there is' has its lightweight or its heavyweight use? The answer to both questions is the same: "attention to context, common sense, and so on."

I will note, finally, that in an earlier version of this statement, I attributed the following view to Craig:

> Existential words and phrases ('. . . exists', 'there is a . . .') have two senses, a lightweight sense and a heavyweight (or "ontologically loaded") sense,

whereas his view is actually this:

> Existential words and phrases ('. . . exists', 'there is a . . .') have two *uses*, a lightweight use and a heavyweight (or "ontologically loaded") use. The meaning or "sense" of those words and phrases is neutral between those two uses.

This section has been purged of that mistake, but some of the things Craig says in his Reply are directed at the mistake in the earlier version and do not apply to the present version of this document. I should say that the corrections I have made were in response to Craig's protests against my misinterpretation in those passages. It is my fault, and not Craig's, that the revised version of this section was not available to him in time for him to revise the criticism of the section in his Reply.

7. The Implications of the Meta-Ontology Presented in Sections 3–5 for the Nominalism-Platonism Debate

At the meeting of the Eastern Division of the American Philosophical Association in New York in 2019, I took part in an author-meets-critics symposium whose subject was *God Over All: Divine Aseity and the Challenge of Platonism*. In my judgment, some of the things Craig said in his reply to my contribution to the symposium provide a better springboard for the discussion of the consequences of the differences between his meta-ontological views and mine than do any passages in the book itself.[19]

One of the things he said in his reply was

> We say things like "There's a lack of compassion in the world," or "There's a better way to do that," without thinking that we are making ontological commitments or need to offer a paraphrase.
>
> (p. 283)

(There are, in this case, passages in the *God Over All* that say essentially the same thing. Indeed 'a lack of compassion in the world' is one member of a long list—compiled by Hofweber—of "things we ordinarily say there are" that is presented on pp. 111–12 of *God Over All*.)

I whole-heartedly assent to this statement. Like Craig, I regard it as a philosophical datum. It will, however, be evident from what I said in Section 5 that he and I would explain this datum and a vast array of similar data in different ways:

19. Both my contribution to the symposium and Craig's reply to his critics have since been published: Van Inwagen (2019); Craig (2019). Unattributed page references in the present section are to those two papers.

CRAIG: When we say things like "There's a lack of compassion in the world" and "There's a good chance that won't happen," we undertake no ontological commitments because we are using 'there's' in an ontologically non-committal way. And the existential quantifier of formal logic ('∃') is neutral between them—equally suited to representing either and favoring neither.

VAN INWAGEN: The syntax of those sentences is a misleading guide to the ontological commitments one undertakes by uttering them assertively. For example, the sentences 'There's a lack of compassion in the world' and 'There are not enough compassionate people and far too many people who are not compassionate' express the same proposition (or as near as makes no matter)— that is, if Alan says, "There's a lack of compassion in the world," and Belinda says, "There are not enough compassionate people and far too many people who are not compassionate," Alan and Belinda have said the same thing in different words. The syntax of the latter sentence is a more reliable guide to the ontological commitments one undertakes by affirming this proposition than is the syntax of the former.[20]

Now let us ask: what are the consequences for the nominalism-realism debate of Craig's way of accounting for these data and of my way of accounting for them? How do these consequences differ?

It will instructive to consider the consequences of these two different approaches to the data for an understanding of a "real" bit of philosophical text.

In Church, 1958 (a classic paper), the twentieth-century American logician and philosopher Alonzo Church quoted several examples of

20. When Alan says, "There's a lack of compassion in the world," one cannot infer from that statement that he believes that something is a lack of something, to wit, of compassion, and that it's in the world. But when Belinda says, "There not enough compassionate people and far too many people who are not compassionate," it would probably be safe to infer that (if she is not dissembling) she believes that there are human beings who are not compassionate.

passages from the works of eminent philosophers[21] who, he maintained, had flatly contradicted themselves in the matter of the existence of abstract objects. One of these examples was from a lecture by one of his contemporaries, the English philosopher A. J. Ayer:

> [I]t makes sense to say, in a case where someone is believing, that there is something that he believes. But it does not follow from this that something must exist to be believed, in the way that something must exist to be eaten or to be struck.[22]

Commenting on these words, Church said:

> This is the passage which shows most clearly that Ayer wishes simultaneously to say there is something that someone believes and to deny that there exists something to be believed.
>
> (p. 1010)

In my initial remarks on *God Over All* in the author-meets-critics symposium, I quoted the passage from Ayer's lecture and Church's comment on it ("trenchant," I called it), and went on to say,

> I will add my own comment, perhaps less trenchant than Church's but very much in the same spirit as his. I should like to have been in a position to ask Ayer whether his words implied the following statement:
>
> > '$\exists x$ is believed by the pope' does not follow from '$\exists x$ the pope believes x'.
>
> And if I had asked that question and Ayer had replied that his words did not imply that statement, I should have asked him how it could *be* that they didn't—for they certainly *seem* to have precisely that implication (p. 274).

And in his response to his critics, Craig said:

> Van Inwagen's claim that Ayer's words involve a glaring apparent contradiction is purchased only at the expense of ignoring Ayer's full claim, which

21. A. J. Ayer, Arthur Pap, and Gilbert Ryle.
22. Ayer (1947). (A pamphlet containing Ayer's inaugural lecture as Grote Professor of the Philosophy of Mind and Logic at the University of London.) I cannot think why Ayer used the bizarre phrase 'a case where someone is believing'—which is hardly English at all. A better sentence would have been, 'It makes sense to say, in a situation in which someone believes that something is the case, that there is something that he believes'.

is not even apparently contradictory. What Ayer said was that "it makes sense to say, in a case where someone is believing, that there is something that he believes. But it does not follow from this that . . . something must exist to be believed, in the way that something must exist to be eaten or to be struck." The first sentence is a lightweight quantificational claim; the second is a heavyweight existence assertion. You get a contradiction only by assuming that the . . . existential quantifier carries heavy ontological commitments, which is exactly what van Inwagen has elsewhere affirmed *but* what neutralists . . . deny (p. 282).

This passage does not seem to me really to reflect an understanding of the point I was making. But if Craig has not understood my point, this may have been at least partly my fault. I may have failed to make what I intended to convey clear owing to the brevity of what I wrote, and, perhaps more importantly, owing to my use of the symbol '∃'. I will present essentially the same point at somewhat greater length and (availing myself of the conceptual machinery developed in the Sections 3–5) without using that contentious symbol:

> The passage would *seem* to commit Ayer to the truth of both the following statements:
>
> 3. If the pope believes that God exists, then there is something that the pope believes.
> 4. There exists nothing that the pope believes.
>
> If Ayer is of the opinion that the pope believes that God exists, then—in virtue of (3)—he (Ayer) ought also to have the following belief:
>
> 5. There is something that the pope believes.
>
> But if Ayer is a participant in a discussion in the Ontology Room, he would be well advised not to use both sentence (4) and sentence (5) as vehicles of assertion. For if one uttered (4) assertively in the Ontology Room, one would thereby express the proposition that
>
> 4′. ~ something^x is such that (the pope believes it^x).
>
> And if one uttered (5) assertively in the Ontology Room, one would thereby express the proposition that
>
> 5′. Something^x is such that (the pope believes it^x).

If we adapted Craig's criticism of the earlier, briefer, '∃'-infected text to this longer, '∃'-free text, we should obtain something like.

> Sentence (4′) is the denial of a heavyweight existence claim, but sentence (5′) is a lightweight quantificational claim. The two sentences are, therefore, not logically inconsistent.

This criticism—obviously enough—is correct only if 'something is such that' has both a lightweight use and a heavyweight use. Can that be so? Craig's reasons for saying that 'exists' and 'there is' have two uses, even if they are valid, do not apply to the phrase 'something is such that'. (Either as a phrase of Revised English or as a phrase of English—for its meaning in Revised English *is* its meaning in English: it is one of the phrases of English that was "left just as it was" by the sequence of modifications by which English was turned into Revised English.) Those reasons pertain to his explanation of the Two Kinds Fact—a fact about sentences containing words and phrases like 'exist' and 'there is'. Granted, if Alan says, "There are philosophers who think the logical argument from evil is unanswerable," and Belinda says, "There are flaws in the logical argument from evil," the case for saying that Alan is ontologically committed to philosophers is much stronger than is the case for saying that Belinda is ontologically committed to "flaws." This is a datum, and it must be accounted for. But there is no such datum in respect of people's use of 'something is such that'—if only because "people" do not use it.[23] (Unless those people are philosophers trying to speak very carefully.) It seems evident to me (at any rate) that there is no analogue of the Two Kinds Fact for 'something is such that' and the sentences of Revised English in which it occurs. I conclude that the passage Church quoted from Ayer has no consistent Revised English translation—or at least none that is evident.

There is, however, more to be said. I have, after all, conceded that some sentences that contain 'there is' and 'exist' are vehicles of ontological commitment and that some aren't. Might it not therefore be that Craig's criticism of my discussion of Ayer's words is

23. In term of Saussure's famous distinction between *langue* and *parole*: *langue* provides speakers with the "option" of utilizing the expression 'something is such that', but, although it exists as a formally possible expression, that possibility is realized only very rarely in *parole*.

essentially right?—right even if 'there is' *et al.* do not have an ontologically neutral use? I said in the closing paragraphs of the previous section that although *operators* like 'there is' and 'there exist' do not have heavyweight and lightweight use, and even if *predicates* like 'exists' do not have heavyweight and lightweight uses, *sentences* in which those operators and predicates occur can have heavyweight and lightweight *readings*. I have, in effect, given both of Ayer's sentences 'There is something that he believes' and 'There exists nothing that he believes' a heavyweight reading. Might it not be that only the latter should receive a heavyweight reading, and that the former should receive a lightweight reading? Perhaps. Note that if we say that we will give a lightweight reading to Ayer's 'There is something that the pope believes', we are saying that we will not regard its syntax as a reliable guide to the way to the syntax of the sentence of Revised English with which it is to be replaced in the Ontology Room.

My problem is this: I do not know what that "replacement sentence" could possibly be. Suppose someone were to make a speech that began like this:

> The syntax of 'There exists nothing that the pope believes' is to be taken seriously. That is, the correct translation of this sentence into Revised English is (4′). But the syntax of 'There is something that the pope believes' is not to be taken seriously. The correct translation of 'There is something that that the pope believes' into Revised English is not (5′) but a sentence with an entirely different syntactical structure. And that sentence is . . .

I am unable to go beyond this point in this imaginary speech, owing to the fact that I have no idea what "that sentence" might be.

So: We have a bit of text that affirms a nominalistic thesis; and we have compared what I (whose meta-ontology is that of Quine and Church) say about with what Craig (whose meta-ontology is that of Azzouni and Hofweber and Adams) says about it. My hope is that this comparison has shown why it is that different meta-ontologies can lead to different answers to the question, "Is nominalism consistent with our everyday assertions and beliefs?"

Let us now return to the thesis that 'there is' *et al.* have both a lightweight and a heavyweight use—the central thesis of the Craig-Azzouni-Hofweber-Adams meta-ontology. (If 'there is' has both a lightweight and a heavyweight use, it seems inevitable to suppose that 'something is such that' has a lightweight and a heavyweight use.) I have argued that one cannot establish this thesis by arguing that it is needed to explain the Two Kinds Fact, owing to there being a plausible alternative explanation of that fact. But to refute an argument for a certain conclusion is not to refute that conclusion. For all a lightweight use has not been proved to exist by that argument, a lightweight use and a heavyweight use of 'exist' might both exist. (Perhaps the lightweight use of 'exist' is twice exemplified in that very sentence!) And the existence of both ways of using 'exist'—even their lightweight existence—is all that is needed to absolve nominalists of the charge of continually saying things inconsistent with nominalism. (I think we must suppose that if either exists, the other does. Philosophers of the Craig-Azzouni-Hofweber-Adams school seem to think that we followers of Quine maintain that the heavyweight way of using 'there is' and 'exist' is the only way. But that gets us wrong: we deny the *duality of use*. We contend that the one sense of 'exist' cannot be meaningfully described by either term. I shall briefly return to this point at the end of the present section.

Let us concentrate on the lightweight use. This is supposed to be a use of 'exist' *et al.* that is "existentially non-committing" or "not existentially loaded." But *is* there a way of using phrases like 'there is' and 'something is' (and '*il y a*', '*es gibt*', . . .) and 'exists' itself that yields existential (or at least apparently existential) statements that can be described as 'existentially non-committing' or 'not existentially loaded'? It is important not to confuse this question with the question

> *Are* there existential (or at least apparently existential) statements that can be described as 'existentially neutral' or 'not existentially loaded'?

The answer to the latter question is, Yes, of course. ('There's a lack of compassion in the world,' for example.) The former question does not ask whether such statements *exist*, but asks rather whether their existence is due to a certain thing, namely that 'there is' and 'exist' *etc.* can be "used in an existentially non-committing way" or "used without existential loading."

What—I ask—is this existential loading, this feature that 'exist' is alleged to have when it is used one way and to lack when it is used in another? What do 'the existentially loaded way of using "exist"' and 'the existentially non-committing way of using "exist"' *mean*?

These are good questions. But there is a prior question, to wit, What does 'exist' mean? What definiens can we find to place to the right of 'x exists $=_{df}$'?

And there is a question that is prior to *that* question: What terms can properly occur in that definiens? For such terms must be terms whose meanings are less philosophically controversial than the meaning of 'exist': there is little to be gained by offering a definition of 'exist' whose definiens contains words whose meaning is as controversial as (or more controversial than) the meaning of 'exist'.

I will make some suggestions.

First, it seems to me that the meaning of 'it is not the case that' is wholly uncontroversial.

(In what follows, I will use the symbol '~' in place of 'it is not the case that'—an abbreviation that will save a great deal of space.)

Secondly, the meaning of the **numerical-identity** sign '=' is at least *almost* as uncontroversial as the meaning of 'it is not the case that'.

Numerical and Descriptive Identity

"Mary and her brother Frank drive the same car." "Yes, Frank liked Mary's 1955 Corvette so much that he bought one himself."

"Mary and her brother Frank drive the same car." "Yes, it's a 1955 Corvette. Their father left it to them."

The first exchange implies that the car Mary drives and the car Frank drives are (in at least one respect) **descriptively identical**—that is, the description '1955 Corvette' applies to them both.

The second exchange implies that the car Mary drives and the car Frank drives are **numerically identical**—that is, there's only *one* of them. Or: when one speaks of the car Mary drives and of the car Frank drives, the number of cars one is talking about is 1.

Logicians use '=' (pronounced "is identical with," not "equals") to express numerical identity.

Some examples of numerical identity:

> Mark Twain = Samuel Clemens (there is a certain man who is the referent of both 'Mark Twain' and 'Samuel Clemens')

Abraham Lincoln = the sixteenth president of the United States (there is a certain man who is the referent of both 'Abraham Lincoln' and 'the sixteenth president of the United States')

π = the ratio of the circumference of a circle to its diameter (there is a certain number that is the referent of both 'π' and 'the ratio of the circumference of a circle to its diameter')

7 + 5 = 12 (there is a certain number that is the referent of both '7 + 5' and '12').

Thirdly, I would suppose that the meaning of 'everything is such that' is less controversial than the meaning of 'exist'. (I intend the words 'everything is such that' to be synonymous with 'everything *whatever*, everything *without exception or qualification* is such that'.)

Those three items will suffice—that is, I contend that 'it exists' can be defined in terms of 'it is not the case that', '=', and 'everything is such that':

It exists $=_{df}$ ~ everything^x is such that (~ it^x = it).

(Or 'x exists $=_{df} \sim\forall y \sim y = x$.' Or, in ordinary English, 'It exists $=_{df}$ not everything isn't it.)

For example, 'Vladimir Putin exists' means '~ everything^x is such that ~ it^x = Vladimir Putin' (or, in ordinary English, 'Not everything isn't Putin'). And that is true. For Putin is identical with Putin, so not everything is non-identical with Putin. And 'Pegasus exists' means '~ everything^x is such that ~ it^x = Pegasus' (or, in ordinary English, 'Not everything isn't Pegasus'). And that is false: examine anything you like, past, present, or future, and you will discover that it isn't Pegasus.

The members of one important philosophical school, the followers of Alexius **Meinong**, will say that this definition fails because (e.g.) it is not the case that everything is not Pegasus—and yet (as everyone will agree) Pegasus does not exist.

Meinonginanism

The Austrian philosopher Alexius Meinong (1853–1920) believed that every thought has an object. Displaying an admirable devotion to consistency, he insisted that if Alice is thinking about Pegasus, then her thought has an object, and that object is, of course, Pegasus. Pegasus, however, does not exist and, indeed, does not have being of any kind. He is nevertheless, in some sense, "there" to be thought about. This position raises a question as obvious as it is serious. "But, Meinong—is not 'being there to be thought about' a kind of being?" Meinong was aware of the seriousness of this question. But the closest he came to answering it was to say, "If one were fond of a paradoxical mode of expression, one might very well say, There are objects of which it is true that there are no such objects." And that answer raises a question of its own, a question equally obvious and equally serious: "And how, Meinong, would one express that thought if one were *not* fond of a paradoxical mode of expression?" No answer has been forthcoming. Or, rather, many answers have been forthcoming, but no satisfactory answers. The proposed answers have one and all been statements that were presented as contradiction-free replacements for 'There are objects of which it is true that there are no such objects' but which in fact contain the same contradiction concealed in a fog of obscure language.

Various present-day philosophers believe that there are things that do not exist—or that there *are* objects of which it is true that there *exist* no such objects. There is such a thing as Pegasus they say. After all, if Alice is thinking about Pegasus, there is something she is thinking about. There *is* such a thing as Pegasus—but he does not *exist*. These philosophers are sometimes called **neo-Meinongians**.

For, they will maintain, it is false that *Pegasus* is not identical with Pegasus—and thus true that not everything is not identical with Pegasus. If the proposed definition were correct, therefore, it would be true that Pegasus existed. And that, of course, is not true. And I, of course, will demur: I will say that Pegasus isn't *there* to be (or not to be) identical with anything.

The Meinongians and I differ in ontology (they say, "Pegasus does not exist; therefore, 'Pegasus' refers to something that does not exist"; I say, "Pegasus does not exist; therefore, 'Pegasus' refers to nothing")—for if my meta-ontology is correct (and, in particular, if my definition of 'exists' is correct), their ontology is wrong. They and I constitute an indisputable example of a ground-floor philosophical disagreement. We differ, and everything that can be said in defense of either of our two positions has been said and both sides in the dispute are familiar with and understand all the arguments the other side has to offer them, and neither side regards the opposing arguments as arguments that should convince anyone.

I will say just one thing about the relation between my meta-ontological views and those of the Meinongians. What I shall say is not meant to be a refutation of **Meinongianism**; I say it simply to make my position on Meinongianism clear. I accept both the following judgments:

Meinong and his school accept this proposition.

> There are objects of which it is true that there are no such objects.[24]

This statement is prima facie self-contradictory. No one has ever offered a way of understanding it that eliminates the prima facie contradiction. Nor can the paleo-Meinongians simply "do without" this statement—for the thesis that (in some sense) "there *are* objects that *are not*" is essential to Meinong's position.

24. [*E*]s *gibt Gegenstände, von denen gilt, daß es dergleichen Gegenstände nicht gibt.* (Meinong, 1904. The quoted words occur on p. 490 of vol. II of Haller, Kindinger, and Chisholm, 1969–1973. (P. 83 of the English translation.) Craig has taken this statement to imply that Meinong accepted the existence (or at any rate the subsistence) of both lightweight and heavyweight quantification: the initial 'es gibt' is ontologically neutral, and the terminal '*es . . . nicht gibt*' ontologically loaded (Craig, 2016, p. 104). But if that is so, why did Meinong prefix this sentence with '*Wer die paradoxe Ausdrucksweise liebt, könnte also ganz wohl sagen:*' ('If one were fond of a paradoxical mode of expression, one might very well say:')—and then leave it to others to explain how the appearance of contradiction might be removed? If Craig were right, Meinong—surely?—would have gone on to say something like this: "But the apparent contradiction vanishes when one realizes that the initial 'es gibt' in this statement is ontologically non-committal, and the terminal 'es . . . nicht gibt' is ontologically loaded."

Neo-Meinongians contend that Pegasus *is* and the Phoenix *is*, but that they nevertheless do not *exist*. But the neo-Meinongians have never said what they mean by 'exists'. That is, they insist that that 'x exists' does not mean 'not everything is not x' or '$\exists y\, y = x$', but they have never actually said what, in their view, 'x exists' *does* mean.

I now return to the topic of the meaning of 'exists'. I have said that 'it exists' means
'~ everything^x is such that (~ it^x = it)'. I will now consider some consequences of this definition.

I contend that the six sentences in the following list all mean the same thing—or at least that no two of them differ in meaning more than do 'All Greeks are mortal' and 'There are no immortal Greeks'.

~ everything is such that (~ it = Descartes)

$\sim\forall x \sim x = \text{Descartes}$

Something is such that (it = Descartes)

$\exists x\, x = \text{Descartes}$

Descartes = Descartes[25]

Descartes exists.

("Are you saying that existence is nothing but self-identity, then?" Yes, I am. I like to quote, in this connection, Auden's great poem "In Memory of W. B. Yeats." Auden describes the afternoon of the poet's death in these words: "it was his last afternoon as himself": that afternoon, someone—namely, Yeats—was Yeats; that evening, no one would be Yeats. Why a Christian poet would say that is, of course, another question.)

And the relations of those six sentences to such more elaborate sentences as

25. Note that (at least in ordinary logic) 'Descartes = Descartes' is equivalent to 'Something is Descartes' (and '$\exists x \exists y\, x = y$' follows from 'Tully = Cicero'). After all, if Descartes is Descartes, then something is Descartes; if something is Descartes—well, that "something" would have to be Descartes, wouldn't it? It certainly couldn't be anything *else*.)

Descartes exists in the strict and philosophical sense of 'exist'

Descartes exists in the metaphysically serious sense of 'exist'

can perhaps be understood by reference to an example we have already used: the relation is the relation of

Henry loves his mother

to

Henry is bound to the maternal author of his being by adamantine chains of filial love.

I contend further that the sentences in each of the following pairs mean the same (more or less; for all intents and purposes):

I think, therefore I am
I think, therefore it is not the case that everything is not I

God exists
It is not the case that everything is not God

She lived unknown, and few could know/When Lucy ceased to be
She lived unknown, and few could know/When it became the case that everything that there then was was not Lucy

It makes me strangely uneasy to contemplate the fact that if my parents had not met I should never have existed
It makes me strangely uneasy to contemplate the fact that if my parents had not met it would always have been the case that everything was not I

Having said what 'exists' means, there is little more I need to say about the question whether 'something is such that' (or any other "existential expression," such as 'there is' and 'exists') has both a lightweight and a heavyweight sense. If my account of the meaning of 'exists' is correct, it obviously makes no sense to say that there is a way of using 'something is such that' that is "not existentially loaded." And this statement should not be taken to imply that I suppose the one use that 'something is such that' has is the so-called heavyweight use: The two ideas (that is, the two "ideas," the two pseudo-concepts) "the lightweight use of an existential expression" and "the heavyweight use of an existential expression" live by taking in each other's washing. *Neither* makes any sense because the supposed opposition that they represent makes no sense.

Craig contends that the existential idioms of ordinary English have both a heavyweight and a lightweight use, a thesis he defends by an appeal to examples: 'There's a statue of Nelson in Trafalgar Square' vs. 'There's fun to be had in Trafalgar Square'. The existential quantifier of formal logic is simply neutral between the two uses, and that is why the existential quantifier cannot be a device of ontological commitment. In reply, it is maintained in Section 6 that the examples do not show that 'there is' *et al.* have a heavyweight and a lightweight use: they rather show that the syntax of an English sentence can be a more or a less reliable guide to the way ordinary sentences containing 'there is' and 'exits' should be translated into a language whose only item of existential vocabulary is 'something is such that'. Section 7 considers (a) some implications of the meta-ontological disagreement that was the topic of Section 6 for ontology (and in particular for the problem of universals), and (b) the question whether 'something is such that' can have both a heavyweight and a lightweight use. It is maintained that they cannot.

8. Substitutional Quantification

One objection to my position remains to be considered. Craig has said,

> There is [a] semantics for quantificational discourse which is admitted on all sides not to be ontologically committing to objects in a domain, namely, so-called substitutional quantification. On a substitutional understanding of quantification, one does not take the variables within the scope of the quantifier to range over a domain of objects; rather, we take the variables as dummy letters which may be replaced by linguistic expressions in order to form sentences. . . An existentially quantified statement is true just in case the substitution of at least one term for a variable in the sentence following the quantifier yields a true sentence (*God Above All*, p. 108).

That is not quite how I would describe **substitutional quantification.** I prefer to say that there are two quantifiers (that is, two universal quantifiers and two existential quantifiers—or perhaps two universal quantifiers, one of which has the existential quantifier as its dual and the other of which has the "particular" quantifier as its dual). The two existential/particular quantifiers are, of course, the existential quantifier (objectual) and the **particular quantifier** (substitutional)—not one operator that can be understood in either of two ways, but two operators each of which is to be understood in one way. I will use the symbol '∃' for the former and the symbol 'Σ' for the latter. As I see matters, the sentence

∃*x* *x* is a dragon

is true just in the case that something is such that it satisfies the open sentence '*x* is a dragon'.[26] (That is to say, 'Something is such that it is a dragon' is true just in the case that something is such that it satisfies the free pronominal sentence 'it is a dragon'.[27]) I make bold to say that natural history reveals to us that it is false that something is such that it satisfies the open sentence '*x* is a dragon'. Therefore, '∃*x* *x* is a dragon' is not true.

By contrast, the sentence

Σ*x* *x* is a dragon

is true just in the case that there is a term such that the result of replacing '*x*' in '*x* is a dragon' with that term is true.[28] Suppose I am

26. I have no use for the idea that a sentence in the quantifier-variable idiom has a truth-value only relative to a "domain of quantification."
27. A free pronominal sentence is a sentence of Revised English that contains occurrences of 'it' that have no antecedent. A thing satisfies a free pronominal sentence just in the case that that sentence is true of it. ('It is human' is true of Socrates, 'it is a dragon' is true of nothing.)
28. The friends of substitutional quantification insist that a Σ-sentence is not a disguised objectual quantification over linguistic items—and is thus not equivalent to the truth-conditions they have provided for it. For example, 'Σ*x* *x* is a dragon' is true if and only if there is a term such that the result of replacing '*x*' in '*x* is a dragon' with that term is true; but 'There is a term such that the result of replacing "*x*" in "*x* is a dragon" with that term is true' is not what that Σ-sentence *means*. See Orenstein (1978).

willing to say that 'Fafner is a dragon' is true.[29] I can then reason as follows. Since 'Fafner' is a term, and 'Fafner is a dragon' is the result of replacing 'x' in 'x is a dragon' with 'Fafner', and 'Fafner is a dragon' is true, 'Σx x is a dragon' is true.

And might not "the ontologically non-committal use" of 'there is' be simply the "substitutional" use?

Might it not be that if we imagine Ayer saying,

> There is something that the pope believes.

we should, if we wish to represent his actual views, imagine that he is so using 'there is' that 'There is something that the pope believes' follows from 'The pope believes that God exists'. And might it not also be if that when we imagine him saying

> There exists nothing that the pope believes,

we should, if we wish to represent his actual views, imagine that he is so using 'there is' that, 'There exists nothing that the pope believes' implies that everything—every object—there is such that it is not an object believed by the pope. There would then be no inconsistency in the pope's beliefs. Well, it's an idea worth thinking about. Let us therefore think about it.

There is, I maintain, an important difference between '$\exists x$ x is a dragon' and 'Σx x is a dragon'—that is, an important difference other than their having different truth-conditions. And the difference is: the former means something and the latter means nothing. Or, at any rate, the friends of substitutional quantification have never *said* what the latter means.

It is not so with '$\exists x$ x is a dragon'. I can tell you what that sentence means, or at least I can tell you what I *say* it means. And that is

> Something^x is such that (it^x is a dragon).

(For '$\exists x$ x is a dragon' is—so *I* say—an abbreviation of the offset sentence: '$\exists x$ x is a dragon' is obtained merely by abbreviating 'Something^x is such that' as '$\exists x$' and '(it^x is a dragon)' by '(x is a dragon)'.

---

29. I am in fact *not* willing to say this. I think that there are contexts in which that sentence would express a true proposition and others in which it would not. I am unwilling therefore to ascribe any truth-value to that sentence per se.

But what does 'Σx x is a dragon' mean? We have not been told. The only thing we have been told about "Σ-sentences" is what their truth-and-satisfaction-conditions are, and two sentences with the same truth-conditions or the same satisfaction-conditions can have different meanings—witness the two sentences '(Lions are carnivores or lions are herbivores) & lions are carnivores' and 'Lions are carnivores', and the two sentences '(x is a carnivore or x is an herbivore) and x is a carnivore' and 'x is a carnivore'.

In Van Inwagen (1981), I presented an argument for the conclusion that to supply truth-and-satisfaction-conditions for Σ-sentences is not to explain what they mean. Since few philosophers if any found the argument of that essay convincing, I will try a different tack—an intuition pump, I suppose one could call it.

Suppose I were to "introduce" a variable-binding operator 'Λ' by saying the following things and nothing more:

> 'Λx if x is Greek, then x is mortal' has the same truth-value as '$\forall x$ if x is Greek, then x is mortal'
>
> 'Λ' means something different from '$\forall$' and 'Λx if x is Greek, then x is mortal' means something different from '$\forall x$ if x is Greek, then x is mortal'
>
> In those two statements, substitute any variable for 'x' and any open sentence containing only the variable substituted for x, and the results will be true. For example, these two sentences are true:
>
> 'Λy y is a material thing' has the same truth-value as '$\forall y$ y is a material thing'
>
> 'Λ' means something different from '$\forall$' and 'Λy y is a material thing' means something different from '$\forall y$ y is a material thing'.

Would you then know what 'Λ' and, e.g., 'Λy y is a material thing' meant? No, you would not. You would have *no idea* what those items meant.[30]

30. You would know that (given materialism) 'Λy y is a material thing' is *true* if it is meaningful (that if it expresses any proposition at all, the proposition it expresses is true), but you would not know what it *meant*: you would not know whether it expressed a proposition, and if some consideration convinced you that it expressed

And *I* have no idea what 'Σx *x* is a dragon' means. And if I have no idea what 'Σx *x* is a dragon' means, I have no idea what *any* Σ-sentence means. And if I have no idea what any Σ-sentence means, neither does anyone else—for no one knows anything about Σ-sentences but what I know: to wit, what their truth-and-satisfaction-conditions are. And if no one has any idea what any Σ-sentence means, no one has any idea what meaning the sentence

> Ayer's sentence 'There is something that the pope believes' means 'Σx the pope believes *x*'

ascribes to 'There is something that the pope believes'.

It has been maintained that, e.g., '$\exists x$ (. . . *x* . . .)' means 'Something[x] is such that (. . . it[x] . . .)'. If that is the case, then the truth-conditions for '$\exists$' sentences are "objectual": '$\exists x$ *x* is a horse' is true just in the case that something—some *object*—is such that it satisfies the condition stated by the open sentence '*x* is a horse'. But suppose we regard the sentence 'Pegasus is a winged horse' as true. We shall probably not want to say that '$\exists x$ *x* is a winged horse' is a logical consequence of 'Pegasus is a winged horse'. But might there be a kind of quantification that has this feature?—that allows us to say things like, "Yes, there is a winged horse in Greek myth—Pegasus, who was captured by Bellerophon." Why not? We use the symbol 'Σ' for it, and we can say that the truth-conditions for 'Σ' sentences are "subsitutional." And then, might it not be that the lightweight use of 'there is' is simply to use 'there is' to express substitutional quantification?

The answer to this question is a simple No—for substitutional quantification is meaningless.

a proposition, you would not know what proposition it was that it expressed. And that's certainly possible: I know (on the basis of testimony) that the sentence 'Satisfactory recursive numerical algorithms exist for summing tree-level Feynman diagrams' expresses a true proposition, but I have no idea what that proposition is. And if I had encountered that sentence only in some "non-authoritative" context—in a science-fiction story, for example—I should have had no idea whether it expressed any proposition at all.

Chapter 2

Opening Statement

William Lane Craig

Contents

1. Introduction

When we ask about the reality of numbers, our question is apt to engender bewilderment. Of course, numbers exist, and here are some of them: 1, 2, 3, 4. But those ink marks are properly called **numerals** rather than **numbers**; we could have used other marks instead, like Roman numerals: I, II, III, IV. Numbers are what numerals stand for, what we refer to by means of numerals. But do numbers themselves exist? Are numbers as real, say, as rocks and people and electrons? Are they actual things that are part of mind-independent reality? Do numbers *really* exist?

Our question epitomizes a broader metaphysical debate about a wide array of putative objects thought to be abstract in nature, including such things as **properties**, **propositions**, **possible worlds**, sets, fictional characters, even literary and musical compositions. Do such abstract objects really exist? In order to avoid confusion in addressing this question, it will be helpful at the outset to adopt a consistent terminology for our discussion.

The first term to need clarification is "**Platonism.**" Plato held that there exist uncreated entities other than God. These are not part of the physical world, which God has created, but are part of a **transcendent**, conceptual realm comprising what Plato called

DOI: 10.4324/9781003008712-3

Ideas or **Forms**. They include mathematical objects like numbers and geometrical shapes, such as the perfect circle or triangle, which are not to be found in the physical realm. Plato held that, far from being created by God, these transcendent realities served as God's model or pattern after which He fashioned the physical world.[1] For Plato, the objects of this transcendent realm are actually *more* real than the objects of the physical world, which are like mere shadows of these transcendent realities. Plato held that these transcendent, ideal objects are uncreated, necessary, and **eternal**.

Contemporary Platonism differs vastly from classical Platonism in various respects[2]; but both views are united in holding that there exist nonphysical, uncreated entities—for example, mathematical objects—other than God. Contemporary Platonists call such entities "abstract" objects in order to distinguish them from concrete objects like people, planets, and chairs.

Today there are actually two very different views both claiming the label of "Platonism." One is a sort of **"heavyweight" Platonism** which takes abstract objects to be just as real as the physical objects that make up the world. As Michael Dummett says, "The mathematician is, therefore, concerned, on this view, with the correct description of a special realm of reality, comparable to the physical realms described by the geographer and the astronomer."[3] For the heavyweight Platonist, our ontological inventory of the world must therefore include numbers along with concrete objects.

By contrast, there is also a sort of **"lightweight" Platonism** whose ontological commitment to abstract objects is much more obscure. For thinkers of this stripe abstract objects seem to be merely semantic objects: they are what we are talking about when we use abstract terms like "3" or "the set of even numbers." They need be no more real than grammatical objects. Something can be grammatically the direct object of a sentence without being a really existing object, as in the sentence "The Press Secretary knew the whereabouts of the

1. See his dialogue *Timaeus* 3–4.
2. Principally, in taking abstract objects to be causally unrelated to the concrete world; neither do contemporary Platonists consider abstract objects to be more real than concrete objects. Nor do they think that concrete objects participate in some way in abstract entities, as Plato thought physical objects participate in ideal objects.
3. Michael Dummett, "Platonism," in *Truth and Other Enigmas* (Cambridge, MA: Harvard University Press, 1978), p. 202.

President." Similarly, the phrase "the whereabouts of the President" can be semantically a term we use to talk about his whereabouts, that is, the term refers to his whereabouts, without implying that there is some really existing object which is the President's whereabouts. These lightweight Platonists—who are among the most ardent defenders of Platonism today—thus seem to be committed to abstract objects only in the sense that they are semantic objects. The focus of our debate is heavyweight Platonism, for this is the only kind of Platonism that adds anything to our ontological inventory. We want to understand what grounds there are for affirming heavyweight Platonism and how one might respond to it.

Next we need to ask how we should understand the distinction drawn by contemporary philosophers between **abstract** and **concrete** objects. Although the distinction is commonplace, it remains a matter of dispute just how to draw that distinction. It is very widely held among philosophers that abstract objects, in contrast to concrete objects, have no causal powers and so are not related to other objects as causes to effects. Moreover, their causal impotence seems to be an **essential** feature of abstract objects. The number 2, for example, does not just *happen* to be causally effete. It seems inconceivable that 2 could possess causal powers. On the other hand, many philosophers have given up the job of finding a criterion to distinguish abstract from concrete objects, choosing instead simply to point to examples which serve as paradigms of each type of object. For example, physical objects are universally taken to be examples of concrete objects. On the other hand, mathematical objects like numbers, functions, and sets are widely regarded as paradigmatic examples of abstract objects. So a discussion of the reality of abstract objects is usually able to proceed on the basis of such shared examples, even in the absence of a clearly enunciated criterion distinguishing abstract from concrete objects. In our discussion, we shall focus on the reality of mathematical objects.

So **Platonism** is the view that abstract objects exist, while **anti-Platonism** is the view that abstract objects do not exist. Sometimes these two views are equated with **realism** and **anti-realism**, respectively, but this equation is misleading. For, as we shall see, there are anti-Platonists who believe in the reality of mathematical objects, but who think that these objects are concrete, not abstract. Thus, some anti-Platonists are realists, and some anti-Platonists are anti-realists about mathematical objects. I shall, therefore, take realism to be any view according to which mathematical objects exist and anti-realism to be any view according to which such objects do not exist. So a realist about mathematical objects might be either a Platonist or an anti-Platonist; but all anti-realists are anti-Platonists, since they think that mathematical objects do not exist.

In the contemporary literature both anti-Platonism and anti-realism have been called **nominalism**, which is very confusing, since anti-Platonism and anti-realism are different views. Moreover, the word "nominalism" is a term which is used in two different philosophical debates to denominate very different positions. The first is the age-old dispute over the existence of universals. In this debate, nominalism is the view that universals do not exist, that everything that exists is a particular. The second debate is a very recent discussion, centered in the philosophy of mathematics, that has arisen only since the publication of the German mathematician Gottlob Frege's *Foundations of Arithmetic* (**1884**). In this debate, the word "nominalism" is often used as a synonym for either anti-Platonism or anti-realism. The confusion is augmented by the fact that a person who is a nominalist in one debate may not be a nominalist in the other debate. For example, in the old dispute over universals one type of nominalism is called class nominalism, according to which similar objects are those included in a certain class. Since classes are abstract objects akin to sets, however, such a thinker is *not* a nominalist in the second debate, but rather a realist and a Platonist! So it is best to avoid the word "nominalism" and to speak instead of anti-realism with respect to mathematical objects.

Terms and Definitions

Platonism: the view that abstract objects exist.
Anti-Platonism: the view that abstract objects do not exist.
Abstract objects: objects that are causally effete; paradigmatically, mathematical objects.
Concrete objects: objects that have causal powers.
Heavyweight Platonism: the view that abstract objects are as real as any concrete object.
Lightweight Platonism: the view that abstract objects are merely semantic objects.
Realism: the view that mathematical objects exist.
Anti-Realism: the view that mathematical objects do not exist.
Nominalism: forget about it!

2. The Indispensability Argument for Platonism

In the contemporary debate over Platonism, there is one argument that predominates: the so-called **Indispensability Argument.** The argument's claim is that we are committed to the reality of abstract objects by many of the statements we take to be true, such as mathematical statements like "1 + 1 = 2." Although the Indispensability Argument for Platonism originated with W. V. O. Quine, a giant of twentieth century philosophy, subsequent discussion has largely overtaken Quine's version of the argument, which was predicated upon a number of idiosyncratic and now widely rejected theses.[4] Today there are a variety of versions of Indispensability Arguments on tap, free of Quine's more controversial theses. Philosopher of mathematics Mark Balaguer nicely epitomizes such arguments as follows:

I. If a simple sentence (i.e., a sentence of the form '*a* is *F*', or '*a* is *R*-related to *b*') is literally true, then the objects that its singular terms denote exist. Likewise, if an existential sentence is literally true, then there exist objects of the relevant kinds; e.g., if 'There is an *F*' is true, then there exist some *F*s.

II. There are literally true simple sentences containing singular terms that refer to things that could only be abstract objects. Likewise, there are literally true existential statements whose existential quantifiers range over things that could only be abstract objects.

III. Therefore, abstract objects exist.[5]

Let's unpack these premises for the sake of those unfamiliar with the terminology.

2.1 Premiss (I): A Criterion of Ontological Commitment

Premiss (I) states a **criterion of ontological commitment.** It does not tell us what exists, but it does claim to tell us what must exist if a

4. See my account of Quine's Indispensability Argument in Chapter 3 of my *God and Abstract Objects* (Cham: Springer, 2017).
5. *Stanford Encyclopedia of Philosophy, s.v.* "Platonism in Metaphysics," by Mark Balaguer.

sentence we assert is to be true. It is intended to reveal to us what our discourse commits us to ontologically. According to Premiss (I), we make ontological commitments in two ways.

2.1.1 Singular Terms

According to the first part of Premiss (I), we make ontological commitments by means of **singular terms.** What are they? Singular terms are words or phrases which are used to single out something. They include **proper names** like "John," "H.M.S. Bounty," "Blue Velvet," and so on; **definite descriptions** like "the man in the gray suit," "your sister-in-law," "my worst nightmare," and so on; and **demonstrative terms** like "this pancake," "that boy," and so on.

In Premiss (I) Balaguer uses "*a*" as a logical constant for which we may substitute a singular term to form a simple sentence. "*F*" stands for any property which we might predicate of the individual picked out by the singular term, and "*R*" any relation in which that individual may be said to stand. For example, the simple sentence "The *Queen Mary II* is a huge oceanliner" has the form *a* is *F*, "*a*" standing for the *Queen Mary II* and "*F*" standing for *a huge oceanliner*. The simple sentence "The *Queen Mary II* docks in Southampton" has the form *a* is *R*-related to *b*, "*a*" standing for the *Queen Mary II* and "*b*" for Southampton, and "*R*" standing for the relation *docks in*. In the following, we may ignore without detriment simple sentences involving relations and focus on those which are predications of a property to an individual.

The claim of Premiss (I) is that if a simple sentence of the form *a* is *F* is literally true, then the object denoted by its singular term exists. In our example above, the term "*Queen Mary II*" denotes the *Queen Mary II* (or, in other words, the *Queen Mary II* is the denotation of the singular term "*Queen Mary II*"). Accordingly, since the sentence "The *Queen Mary II* is a huge oceanliner" is literally true, the *Queen Mary II* exists.

Notice Balaguer's qualification that the sentence must be ***literally* true.** Sentences employing metaphors and other figures of speech may be true, but they are not literally true. It may be true, for example, that "It's raining cats and dogs!", but it would be obtuse to think that someone asserting such a sentence believes that there are animals falling from the sky. He simply means that it's raining hard. Singular terms employed in non-literal speech are not ontologically committing for their user. But Premiss (I) states that if a simple sentence with a singular term is literally true, there must exist an object corresponding to the singular term used in the sentence.

Why *simple* sentences? What Balaguer is trying to avoid is not complexity but sentences involving what are called **intensional contexts.** Intensional contexts are non-extensional contexts. **Extensional contexts** are sentence phrases which have two characteristics: (i) Singular terms referring to the same **entity** can be switched without affecting the sentence's truth value. For example, "The morning star is Venus" is an extensional context because one could substitute the co-referring term "the evening star" for "the morning star" without affecting the sentence's truth value. By contrast "Ancient Babylonian astronomers believed that the morning star rises in the morning" involves an intensional context because substitution of "the evening star" for the co-referring term "the morning star" would yield a false sentence. (ii) One can quantify into such contexts from outside the context. For example, "Mars has two moons" permits us to infer, "There is something which is a moon of Mars." By contrast "Le Verrier sought to discover Vulcan between Mercury and the Sun" involves an intensional context because we cannot infer that "There is something that Le Verrier sought to discover."

The criterion of ontological commitment proposed by (I) does not apply when it comes to intensional contexts. For example, use of the singular term "the bogeyman" in the true statement "Johnny fears the bogeyman" does not commit us to the reality of the bogeyman. The inapplicability of (I)'s criterion of ontological commitment to intensional contexts implies that it will be useless for vast stretches of human discourse, for **intentional** attitudes (like fearing, hoping, believing), **modal operators** (like "necessarily, . . ." or "possibly, . . ."), and **temporal operators** (like "it was the case that. . ." or "it will be the case that. . .") all establish intensional contexts.[6] The criterion applies only to sentences which are extensional.

2.1.2 *Existential Quantification*

Singular terms are not the only devices of ontological commitment, according to (I). The second part of (I) claims that we also make ontological commitments by means of existential sentences. What

6. For example, "Possibly, there are pink unicorns" does not permit us to infer that there are possible pink unicorns, and "There were carnivorous dinosaurs" does not permit the inference that there are dinosaurs that were carnivorous.

are they? As Balaguer makes clear in the second part of Premiss (II), he is talking about sentences which involve so-called **existential** (or particular) **quantifiers.** In contrast to **universally quantified statements,** which are true with respect to *all* the members of the domain of **quantification, existentially quantified statements** are true with respect to *some* of the members of the domain of quantification. For example, if our domain of quantification is bears, then the statement "Some bears live near the North Pole" is true just in case at least one member of the domain lives near the North Pole.

In ordinary English, there are a variety of **informal existential quantifiers,** such as "some," "at least one," "there is/are," and "there exists." All of these informal expressions are captured in formal logic by the **formal quantifier** "∃." A sentence like "Some bears live near the North Pole" has the form (∃*x*) (*x* is a bear & *x* lives near the North Pole), which is to be read "There is some *x* such that *x* is a bear and *x* lives near the North Pole." In order for the sentence to be true there must be at least one thing in the domain that can be the **value** of the **variable** *x*. Balaguer is evidently talking about first-order logic, where the variables *x*, *y*, *z* take individual things as their values.

Now the claim of Premiss (I) is that the literal truth of simple sentences involving existential quantification of the form "There is an *F*" commits us to the reality of the object which is an *F*. Here "*F*" stands for a general term like "a man," "buffalo," "facts which science has yet to discover," and so on. Such sentences can be symbolized as (∃*x*) (*x* is an *F*). The existential quantifier "∃" is claimed to express existence, not in a metaphysically lightweight sense, but rather in a metaphysically heavyweight sense. It is a device for making ontological commitments. Thus, the person who makes assertions involving informal quantifiers like "some" and "there is/are" thereby commits himself to the reality of the things in the domain of quantification which, as Balaguer puts it in Premiss (II), his quantifiers "range over," that is to say, the things which are the values of the variables in sentences he regards as true.

Summary

So, in sum, what Premiss (I) expresses is not an ontological claim about what exists, but a meta-ontological claim about how a person commits himself ontologically. The claim is that singular terms and existential quantification are devices of ontological commitment.

2.2 *Premiss (II): Abstract Objects*

Turn now to Premiss (II). It claims that the **denotations** or **referents** of certain singular terms in literally true, simple sentences, for example, "2 + 2 = 4," cannot plausibly be taken to be concrete objects of any kind. For example, the referents of such mathematical terms as "2+2" and "4" are abstract objects. Premiss (II) also excludes taking mathematical discourse to be some sort of figurative language, not to be taken literally. It states that at least some abstract discourse is literally true and therefore commits its user to the reality of abstract objects.

In the second part of Premiss (II), it is likewise claimed that there are literally true existentially quantified statements involving quantification over abstract objects. For example, "There is a prime number between 2 and 4" or "There are prime numbers greater than 100." Again, such statements are not to be construed metaphorically, along the lines of "There's a bee in her bonnet." So existentially quantified abstract discourse also commits its user to the reality of abstract objects.

From (I) and (II), the conclusion (III) follows that abstract objects exist. So the anti-realist must defeat either (I) or (II).

Summary

In sum, Premiss (II) claims that there are literally true, simple sentences involving expressions which refer to and **quantify over** abstract objects. Given that such expressions are ontologically committing (Premiss (I)), anyone who asserts such truths finds himself committed to the reality of abstract objects.

3. Responses to the Indispensability Argument

If we take mathematical objects as our paradigm case, then Figure 2.1 portrays some of the many options concerning their reality, most of which serve as responses to the Indispensability Argument.

The various options can be classed as **realist** (mathematical objects exist); **arealist** (there is no fact of the matter concerning the existence of mathematical objects); or **anti-realist** (mathematical

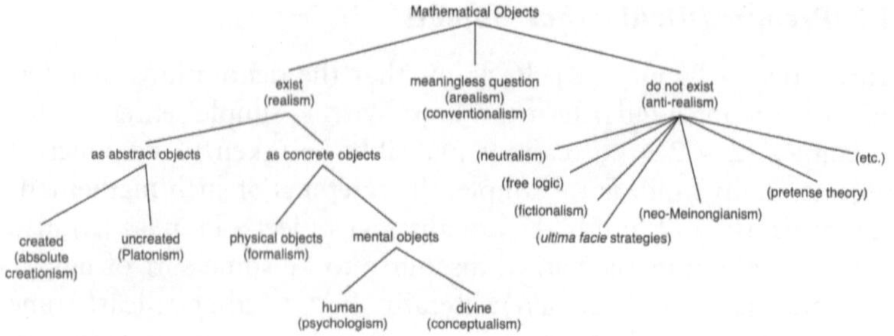

Figure 2.1 Some of the options concerning the existence of mathematical objects.

objects do not exist). As Figure 2.1 illustrates, there are two brands of realism about mathematical objects: views which take them to be *abstract* objects and views which take them to be *concrete* objects. Of realist views which consider mathematical objects to be abstract, there is in addition to Platonism a sort of modified Platonism called **absolute creationism**, which holds that mathematical objects have, like concrete objects, been created by God.[7] Since they are both realists, Platonists and absolute creationists alike may find indispensability arguments unobjectionable. Concretist versions of realism are anti-Platonist views which take mathematical objects to be either *physical* objects or *mental* objects, the latter either in *human* minds or in *God's* mind. The most promising concretist view is some sort of divine **conceptualism**, according to which putatively abstract objects are thoughts in the mind of God.

In between realism and anti-realism about mathematical objects is **arealism**, the view that there just is no fact of the matter about the reality of mathematical objects. The classic version of arealism was the **conventionalism** of Rudolf Carnap.[8] He drew a fundamental distinction between what he called "**internal questions**" and "**external questions**," that is to say, questions which

7. See, e.g., Thomas V. Morris and Christopher Menzel, "Absolute Creation," *American Philosophical Quarterly* 23 (1986): 353–62. For exposition and assessment, see Chapter 4 of my *God Over All: Divine Aseity and the Challenge of Platonism* (Oxford: Oxford University Press, 2016); or Chapter 4 of my *God and Abstract Objects*.
8. Rudolf Carnap, *"Meaning and Necessity:" A Study in Semantics and Modal Logic* (Chicago, IL: University of Chicago Press, 1956), pp. 206–17.

are posed *within* an adopted **linguistic framework** and questions posed by someone *outside* that framework. As a logical positivist and verificationist, Carnap was convinced that metaphysical questions posed outside a linguistic framework have no cognitive content. Whether one adopts a linguistic framework in which it makes sense to speak of numbers is just a matter of convenience, or convention; hence, the name conventionalism. Within that framework, no one would think to deny the existence of numbers; but outside that framework the question is metaphysical and so neither true nor false.

When we turn to anti-realist options, we find a cornucopia of different views. **Neutralism** rejects the criterion of ontological commitment expressed in Premiss (I), taking the use of singular terms and existential quantification to be neutral with respect to ontological commitments. **Free logic**, on the other hand, takes existential quantification to be ontologically committing but denies that the use of singular terms is a device of ontological commitment. **Neo-Meinongianism** denies that existential (or particular) quantification is ontologically committing, though it agrees that there are objects referred to by abstract singular terms—with the caveat that these objects are nonexistent. **Fictionalism** accepts entirely the Platonist's criterion of ontological commitment but denies Premiss (II) that mathematical statements are true. ***Ultima facie* strategies** hold that we can offer **paraphrases** or interpretations of mathematical statements which will preserve their truth value but without their prima facie ontological commitment to abstract objects. **Pretense theory** considers mathematical discourse to be a species of make-believe, so that mathematical objects are akin to entities in fictional stories. And so on!

Summary

There are a wide variety of anti-Platonic views on offer today. Non-Platonic realist views take mathematical objects to be either physical or mental objects. Anti-Realist views dispute either that the use of singular terms and existential quantification in true statements is ontologically committing or that mathematical sentences need be taken to be literally true.

4. Assessment of Anti-Platonist Options

4.1 Responses to Premiss (II)

Consider, first, options which dispute Premiss (II). Realist versions of anti-Platonism dispute that in true mathematical sentences singular terms refer to and existential quantifiers range over things that could *only* be abstract objects. Non-Platonic realists hold that various objects normally thought to be abstract, such as mathematical objects, are in fact concrete. These may be taken to be either physical objects, such as marks on paper, or mental objects or thoughts, either in human minds or in God's mind.[9] Gottlob Frege subjected the views that mathematical objects are physical objects or human thoughts to such withering criticism, however, that such views are scarcely taken seriously today.[10] But Frege's objections to **formalism** and **psychologism**—such as the intersubjectivity, necessity, and plenitude of mathematical objects—do not touch divine conceptualism. With the late twentieth century renaissance of Christian philosophy divine conceptualism is once more finding articulate defenders.

4.1.1 Divine Conceptualism

4.1.1.1 Exposition

Alvin Plantinga, the most influential theist philosopher writing today, has endorsed **divine conceptualism** with regard to supposedly abstract objects. He locates himself in the Augustinian tradition "in thinking of numbers, properties, propositions and the rest of the Platonic host as *divine ideas*."[11] Two Oxonian philosophers Brian Leftow and Greg Welty have provided extensive defenses of this position.[12] Divine

9. For a physicalist view see, e.g., James Franklin, *An Aristotelian Realist Philosophy of Mathematics: Mathematics as the Science of Quantity and Structure* (Houndmills, Basingstoke, Hampshire: Palgrave Macmillan, 2014). I have not been able to find a contemporary proponent of psychologism, the view that mathematical objects are thoughts in human minds. For exposition and assessment, see Chapter 5 of my *God and Abstract Objects.*
10. Gottlob Frege, *The Foundations of Arithmetic: A Logico-Mathematical Enquiry into the Concept of Number*, trans. J. L. Austin, 2d rev. ed. (Evanston, IL: Northwestern University Press, 1968), §I. 7, pp. 8–11; §II. 26–7, pp. 34–8.
11. Alvin Plantinga, "Response to William Lane Craig's review of *Where the Conflict Really Lies*," *Philosophia Christi* 15 (2013): 178.
12. Brian Leftow, *God and Necessity* (Oxford: Oxford University Press, 2012); Greg Welty, "Theistic Conceptual Realism: The Case for Interpreting Abstract Objects as

conceptualism would therefore seem to be an attractive option for any theist who feels the force of arguments for realism.

4.1.1.2 Assessment

Unfortunately, conceptualism is not worry-free. Conceptualism requires that in virtue of His omniscience, God be constantly entertaining actual thoughts corresponding to every proposition and every state of affairs. This may be problematic for the theist. Graham Oppy complains that "it threatens to lead to the attribution to God of inappropriate thoughts: bawdy thoughts, banal thoughts, malicious thoughts, silly thoughts, and so forth."[13] As a theist myself, I take this worry very seriously. If God has the full range of thoughts that we do, then He must imagine Himself, as well as everyone else, to be engaged in bawdy and malicious acts, and, moreover, rather than putting such detestable thoughts immediately out of mind as we try to do, He keeps on thinking about them. What about banal and silly thoughts? Why in the world should we think that God is constantly thinking the non-denumerable infinity of banal and silly propositions or states of affairs that we can imagine? Take Welty's own illustration of the thought that for any real number *r*, *r* is distinct from the Taj Mahal. Why would God retain such inanities constantly in consciousness? Or consider false propositions like *for any real number* r, r *is identical to the Taj Mahal.* Why would God hold such a silly thought constantly in consciousness, knowing it to be false? Obviously, the concern is not that God would be incapable of keeping such a non-denumerable infinity of thoughts ever in consciousness, but rather why He would dwell on such trivialities. It is a *non sequitur* to infer that everything God knows, he is actually thinking about.

Yet another worry for conceptualism is that concrete objects like God's thoughts do not seem suitable to play the roles normally ascribed to abstract objects. Consider **properties,** for example. The chief rationale behind construing properties as abstract universals rather than particulars is the supposed need for an entity that can be wholly located in diverse places. The difficulty, then, for conceptualism is that God's thoughts, as concrete objects, are

Divine Ideas" (Oxford University: Doctoral Thesis, 2006); cf. Greg Welty, "Theistic Conceptual Realism," in *Beyond the Control of God? Six Views on the Problem of God and Abstract Objects,* ed. Paul Gould (Bloomsbury, 2014), pp. 81–111.

13. Graham Oppy, "Response to Welty," in *Beyond the Control of God?,* p. 105.

not universals but particulars and so cannot be wholly present in spatially separated objects. Perhaps in response the conceptualist could say of divine thoughts what the Platonist says of abstract universals: particulars stand in some sort of relation to them in virtue of which particulars are the way they are. An immediate problem for the conceptualist is that if properties are God's thoughts, then particulars must exemplify God's thoughts. But a concrete object does not seem to be the sort of thing that is exemplifiable any more than it can be a universal, since concrete objects are particulars and particulars are not exemplifiable but rather exemplify.

Or consider the suitability of divine thoughts for playing the role of mathematical sets. Plantinga, springboarding off a comment by set theorist Hao Wang about sets' being formed by the mental activity of collecting, suggests that sets be taken to be God's mental collectings. But if sets are really particular divine thoughts, then how do we have any access to sets? The question here is not whether I have a causal connection with sets. Rather it is that sets, the real sets, are locked away in God's private consciousness, so that what we talk about and work with are not sets at all. When I collect into a unity all the pens on my desk, that set is not identical, it seems, with the set constituted by God's collecting activity. Since we have two mental collectings and since sets are God's particular collectings, the "set" I form is not identical to the set of all the pens on my desk. But if sets are determined by membership, how could they not be identical, since they have the same members?

While by no means knock-down objections to conceptualism, these worries should motivate the theist, as well as the non-theist, to look seriously at the wide variety of anti-realist solutions to the challenge of Platonism before acquiescing too easily to the realist viewpoint.

4.1.2 Fictionalism

4.1.2.1 Exposition

Perhaps the boldest option in response to Premiss (II) is **fictionalism.** Fictionalists flatly deny that mathematical sentences are true, period. Statements involving quantification over or reference to abstract objects are false (or at least untrue). Abstract objects are merely useful fictions; that is to say, even though no such objects exist, it is useful to talk as though they did. Hence, the name fictionalism.

Physical science is up to its neck in mathematical statements. What would be the consequences for science of taking mathematical sentences to be untrue? Balaguer thinks that the falsity of mathematical sentences does not undermine natural science because what he calls the non-Platonistic content of scientific theories is independent of and, hence, unaffected by the falsity of their Platonistic, mathematical content. Balaguer is willing to concede claims about the indispensability of mathematics to physical science, but he maintains that while the non-Platonistic content of empirical science is (mostly) true, its Platonistic content is not. Empirical theories use statements about mathematical objects only instrumentally as descriptive aids in order to make assertions about the physical world. Because abstract objects are causally unconnected to the physical world, there must be a non-Platonistic content of scientific theories, which, even if inexpressible by us due to mathematics' indispensability, is made true by the physical world wholly independently of whether abstract objects exist. Thus, says Balaguer, the claim that empirical science has a non-Platonistic content that captures its complete picture of the physical world "is no more controversial than the claim that abstract objects (if there are such things) are causally inert."[14]

4.1.2.2 Assessment

The most evident objection to fictionalism is that some mathematical statements, like "2 + 2 = 4," are just obviously true. Indeed, they seem to be necessarily true. Hartry Field reacts to the claim that a mathematical assertion like "2 + 2 = 4" must be true simply as a consequence of the meaning of its terms by saying that this claim cannot be right because analytic truths cannot have existential implications. He grants that

> the claim 'If there are numbers then 2 + 2 = 4' has some claims to count as an analytic truth, indeed one so obvious that its denial is unintelligible. . . . But. . . it can't be an analytic or purely conceptual truth that there *are* objects 1, 2, 3, 4 etc. obeying such laws as that 2 + 2 = 4. An investigation of conceptual linkages can reveal conditions that things must satisfy if

14. Mark Balaguer, *Platonism and Anti-Platonism in Mathematics* (New York, NY: Oxford University Press, 1998), p. 135.

> they are to fall under our concepts; but it can't yield that there are things that satisfy those concepts. . . .[15]

Field rightly contends that we cannot infer the existence of things from merely definitional truths. But it is only the unquestioned presupposition of the **neo-Quinean** criterion of **ontological commitment** that leads the fictionalist to think that a statement like "2 + 2 = 4" has existential implications. Deny that criterion and one is not forced into the awkward position of denying that 2 + 2 = 4.

In short, one's attitude toward the objection from the obviousness of elementary arithmetic is probably going to depend on one's attitude toward the neo-Quinean criterion of ontological commitment. If with the fictionalist we are convinced that quantification and singular terms are devices of ontological commitment, then we shall find upon reflection that the sentences of elementary mathematics, if taken literally, are anything but obvious. For we shall come to see that statements which we have unhesitatingly accepted as true since childhood are, in fact, radical ontological assertions about the existence of mind-independent abstract objects. As such, they are not at all obviously true. We come to realize that we have, in fact, misunderstood them all these years; we literally did not understand what we were asserting.

On the other hand, if we find sentences of elementary arithmetic to be obvious because we do not take them to be ontologically committing, then we shall be led to reject the neo-Quinean criterion of ontological commitment which would saddle us with such commitments. This seems to me the obvious course to take. Indeed, the anti-Platonist has a quick and easy argument to that end:

1. If the neo-Quinean criterion of ontological commitment is correct, then $2 + 2 \neq 4$.
2. $2 + 2 = 4$.
3. Therefore, neo-Quinean criterion of ontological commitment is not correct.

After all, the sentences of elementary mathematics are much more obviously true than any criterion of ontological commitment and so

15. Hartry Field, *Science without Numbers: A Defence of Nominalism* (Princeton, NJ: Princeton University Press, 1980), p. 5.

should be more tenaciously held and less quickly surrendered than the neo-Quinean criterion of ontological commitment.

4.1.3 Ultima Facie Strategies

4.1.3.1 Exposition

What I call ***ultima facie*** **interpretive strategies** is a diverse group of anti-realist responses to indispensability arguments united by the conviction that mathematical sentences, even if involving prima facie singular reference to or quantification over abstract objects, are capable of being reformulated or interpreted without prejudice to their truth in such a way as to avoid any ontological commitments to abstract objects. Neo-Quineans themselves do not take us to be committed ontologically to all the things referred to or quantified over in ordinary language. For example, if I say, "There are deep differences between Republicans and Democrats," I do not mean to commit myself to the existence of objects called *differences* (some of which are deep), despite my use of the informal quantifier "there is." In a case like this, I can reformulate my original claim so as to serve the same purpose without such ontological commitments on my part, e.g., "Republicans differ deeply from Democrats." Neo-Quineans demand such paraphrases of mathematical sentences from anti-realists, confident that they cannot be provided.

In fact, however, there are a number of strategies for reformulating or interpreting mathematical sentences which preserve their truth without ontological commitment to mathematical objects. The strategies of modal structuralism, constructibilism, and figuralism come to mind.

4.1.3.1.1 MODAL STRUCTURALISM

Geoffrey Hellman's **modal structuralism** proposes to reinterpret categorical mathematical statements like "2 + 2 = 4" as subjunctive conditionals to the effect that if a certain type of structure *were* to exist, then the categorical statement *would* hold with respect to it. Hellman's modal structuralism is a marriage of a structuralist interpretation of Peano arithmetic, real analysis, and Zermelo-Fraenkel set theory with second-order modal logic. The goal is to "translate ordinary sentences of number theory (or analysis)" and set theory "so that, on the interpretation, they say what *would* be the case in any (arbitrary) structure of the appropriate type without literally quantifying over any objects at all."[16] Modal structuralism, sometimes called **eliminative structuralism**, substitutes for customary existence axioms axioms asserting the possibility of structures of a certain kind, where the possibility in question is taken to be a primitive notion. It is thus, as Hellman wryly puts it, structuralism without structures. Hellman acknowledges that "if one simply reads ordinary mathematical discourse literally, i.e. takes it 'at face value,' one arrives at a Platonist interpretation."[17] But "we seek an alternative, non-literal interpretation of mathematical discourse which can be understood as realistic but in which ordinary quantification over abstract objects is eliminated entirely."[18]

4.1.3.1.2 CONSTRUCTIBILISM

Charles Chihara pursues a different *ultima facie* strategy known as **constructibilism** for defeating indispensability arguments. Like the fictionalist, Chihara thinks that on a literal, prima facie interpretation sentences involving quantification over or singular terms for abstract, mathematical objects are false.[19] But unlike the fictionalist, Chihara sees no reason for adopting such a prima facie interpretation; indeed, quite the contrary, since many mathematical sentences are often known to be true. Chihara therefore offers constructibilism as an *ultima facie* way of reading mathematical sentences that enables us to preserve their truth without embracing the implausible ontological commitments of the prima facie reading. Chihara does not espouse constructibility theory as his final theory of mathematics—he actually prefers a sort of modal structuralism—but he puts it forward merely as an available option whose viability serves to undercut Platonist indispensability claims.[20]

16. Geoffrey Hellman, *Mathematics without Numbers* (Oxford: Clarendon Press, 1989), p. 15.
17. Ibid.
18. Ibid., p. 2.
19. Chihara declares, "I do not think that the axioms of set theory, as standardly (Platonically) understood are literally true assertions . . . . I reject set theory as false when it is literally and Platonically construed" (Charles S. Chihara, *A Structural Account of Mathematics* [Oxford: Clarendon Press, 2004], pp. 205, 290).
20. Charles Chihara, *Ontology and the Vicious Circle Principle* (Ithaca, NY: Cornell University Press, 1973), p. 208; idem, *The Worlds of Possibility: Modal*

Chihara's constructibilism supplements first-order logical quantifiers with what he calls constructibility quantifiers. First-order quantifiers binding variables which have mathematical objects as their values will be replaced by constructibility quantifiers carrying no ontological commitment. Corresponding to the existential and universal quantifiers of standard, first-order logic, constructibility theory utilizes two additional quantifiers ranging over variables whose value will be an open sentence token, that is to say, a sentence token having a free variable, such as "x is the book on the table." Constructibilism is thus a theory about what open sentences are constructible and how these sentences would be related to one another. Constructibilism differs from mathematical constructivism in that the quantifiers are modal: it is not said that any such sentences are constructed but merely that it is possible to construct such sentences. The quantifiers therefore carry no ontological commitments to the reality of open sentences. Constructibility theory is Chihara's answer to Quine's challenge to the anti-realist "to develop a nontrivial system of mathematics for the natural sciences that would not be ontologically committed to mathematical objects."[21] Chihara takes himself to have met that challenge, thereby undercutting the Indispensability Argument for Platonism.

It is noteworthy that the mathematical adequacy of the paraphrases of mathematical sentences offered by these *ultima facie* strategists has gone largely unchallenged. Even Platonists acknowledge that Hellman's modal structuralism and Chihara's constructibilism will each recover in its own idiom the same mathematical truths expressed Platonistically in customary mathematical discourse. In accord with the Quinean metaontological rules for settling ontological disputes, Hellman and Chihara have provided paraphrases of accepted mathematical truths which are ontologically neutral.

4.1.3.1.3 FIGURALISM

Stephen Yablo has coined the word "**figuralism**" for yet another *ultima facie* strategy, according to which mathematical sentences

Realism and the Semantics of Modal Logic (Oxford: Clarendon Press, 1998), p. 310; idem, *Structural Account of Mathematics*, pp. 162, 168; idem, "Nominalism," in *The Oxford Handbook of Philosophy of Mathematics and Logic*, ed. Stewart Shapiro (Oxford: Oxford University Press, 2005), p. 498.

21. Chihara, *Structural Account of Mathematics*, p. 162; *cf.* p. 166.

should not be understood literally but are a case of figurative language. Figurative speech, properly interpreted, may be true even if, taken literally, it is false. For in figurative speech, such as understatement, hyperbole, metonymy, and metaphor, the literal content is not what the speaker is asserting.[22] The distinctive contribution of figuralism to the current debate is its contention that talk of abstract objects ought to be understood figuratively, not literally, and may therefore, *pace* the fictionalist, be taken to be true without ontological commitment to abstract objects. If mathematical language is figurative, then it will be maladroit to ask after the ontological commitments of such discourse construed literally.

Yablo thinks that talk of abstract objects involves the use of "existential metaphors," that is to say, metaphors "making play with a special sort of object to which the speaker is not [ontologically] committed."[23] Yablo draws a number of interesting parallels between talk of Platonic objects[24] and figurative talk (construed as involving make-believe objects).[25] These parallels serve as evidence that abstract object talk is a kind of figurative language. Yablo maintains that the metaphorical or real content of mathematical truths is logical truths, which require no ontological foundation: "Arithmetic is, at the level of real content, a body of logical truths—specifically, logical truths about cardinality—while set theory consists at the level of real content, of logical truths of a

22. Stephen Yablo, "A Paradox of Existence," in *Empty Names, Fiction, and the Puzzles of Non-Existence*, ed. Anthony Everett and Thomas Hofweber (Stanford, CA: Center for the Study of Language and Information, 2000), p. 291.
23. Ibid., p. 293. *Cf.* idem, "Does Ontology Rest on a Mistake?" *Proceedings of the Aristotelian Society* (Supplement) 72 (1998): 250; idem, "The Myth of the Seven," in *Fictionalism in Metaphysics,* ed. Mark Eli Kalderon (Oxford: Clarendon Press, 2005), p. 98.
24. It should be noted that Yablo has an idiosyncratic understanding of what a Platonic object is. Objects are Platonic relative to an area of discourse iff the discourse depends on how those objects behave even though the discourse is not really about those objects. For example, someone who expresses concern about the number of starving people in the world is concerned about people, not some abstract object.
25. Yablo, "Paradox of Existence," pp. 302–4; *cf.* idem, "Go Figure: A Path through Fictionalism," in *Figurative Language,* ed. Peter A. French and Howard K. Wettstein, *Midwest Studies in Philosophy* 25 (Oxford: Blackwell, 2001), pp. 89–90; idem, "Abstract Objects: A Case Study," in *Realism and Relativism,* ed. Ernest Sosa and Enrique Villaneva, *Philosophical Issues* 12 (Boston, MA: Blackwell, 2002), pp. 227–30.

combinatorial nature."[26] Though, there being no numbers, the literal content of arithmetical truths is false, nevertheless their real content is true by definition. The real content of arithmetical truths like 2 + 3 = 5 is the first-order logical truth $(\exists_2 xFx \,\&\, \exists_3 yGy \,\&\, \neg\exists z(Fz \,\&\, Gz)) \rightarrow \exists_5 u(Fu \vee Gu)$.

4.1.3.2 Assessment

So what complaint is there, from the Platonist's perspective, with various *ultima facie* strategies? The most common complaint in the literature is that such strategies either grossly misrepresent mathematical discourse or advocate without sufficient justification a radical revision of mathematical discourse. This complaint is rooted in a familiar distinction drawn by John Burgess and Gideon Rosen between what they call **hermeneutic nominalism** and **revolutionary nominalism.**[27] Hermeneutic nominalism aspires to offer an exegesis of mathematical discourse with a view toward determining what people, especially mathematicians and scientists, themselves mean by their mathematical assertions. The hermeneutic nominalist claims that people do not really take their assertions as prima facie and literally true and offers his *ultima facie* interpretation as the intended content of their assertions. The revolutionary nominalist, by contrast, grants that the prima facie interpretation of mathematical discourse gives the intended meaning of such discourse on the part of its users but offers in its place a novel reinterpretation of such discourse which is not Platonistic. The revolutionary nominalist advocates his favored *ultima facie* interpretation as providing the right interpretation of mathematical discourse in place of the customary interpretation. Given this distinction, which is taken to be jointly exhaustive, the objection is that neither hermeneutic nor revolutionary nominalism is plausible, and therefore both should be rejected. The implication of this objection for *ultima facie* strategies

26. Yablo, "Myth of the Seven," p. 99; idem, "Abstract Objects," pp. 230–32.
27. See John P. Burgess, "Why I Am Not a Nominalist," *Notre Dame Journal of Formal Logic* 24 (1983): 93–105; idem, "Mathematics and Bleak House," *Philosophia Mathematica* 12 (2004): 23–30; John P. Burgess and Gideon Rosen, *A Subject with No Object* (Oxford: Clarendon Press, 1997), pp. 6–7, 205–37; idem, "Nominalism Reconsidered," in *The Oxford Handbook of Philosophy of Mathematics and Logic*, ed. Stewart Shapiro (Oxford: Oxford University Press, 2005), pp. 515–28.

is that the proffered reinterpretations should be rejected in favor of the prima facie interpretation, which is taken to be the default interpretation.

None of the thinkers we have surveyed, however, espouses revolutionary nominalism. On the contrary, they see no reason that mathematicians should not continue in the use of their Platonistic language. Rather they are offering merely an undercutting defeater of Premiss (II) of the Indispensability Argument that mathematical sentences cannot be reformulated or interpreted so as to be truth-preserving but without ontological commitment to abstract objects. Neither are Hellman and Chihara offering a hermeneutical claim, since they say they have no idea what the majority of mathematicians and scientists think about these questions. While Yablo does espouse figuralism as a hermeneutic thesis, there is no reason that the anti-realist has to present it as such. Like Chihara and Hellman, he can remain agnostic, in the absence of linguistic and sociological studies, about hermeneutical questions and present the figurative interpretation as one reasonable way of understanding abstract object talk without commitment to abstract objects. If such an interpretation is reasonable, then the Indispensability Argument has been defeated.

4.1.4 Pretense Theory

4.1.4.1 Exposition

Closely related to fictionalism and figuralism about abstract discourse is what we may call **pretense theory**. According to this view, abstract object discourse is a kind of fictional discourse and thus analyzable in terms of theories of fiction. Like fictionalists and *ultima facie* strategists, pretense theorists do not challenge the neo-Quinean criterion of ontological commitment that comes to expression in Premiss (I) of the Indispensability Argument but instead challenge Premiss (II). Pretense theory differs from fictionalism in that it does not claim that sentences involving quantification over or reference to abstract objects are untrue. But neither does it affirm with figuralism that such discourse is often true. Pretense theory is just neutral about the truth value of such discourse.

The pretense theorist's essential point, rather, is that whether or not such sentences are true, we are invited to *imagine* that they are true. The pretense theorist agrees with the figuralist against the fictionalist that abstract object discourse ought not, or need not, to be taken literally; but he differs from the figuralist in attributing non-literality, not to the figures of speech employed, but to the fictional nature of such discourse. Thus, abstract discourse, being fictional, does not commit us to the reality of abstract objects.

Contemporary theories of fiction draw much of their inspiration from the brilliant, pioneering work of Kendall Walton.[28] Prescribed imagining lies at the heart of Walton's theory of fiction. Fictional propositions are propositions which in certain contexts we are to imagine to be true.[29] Walton recognizes that some propositions which are fictional may not, strictly speaking, be prescribed because they are trivial or distracting (for example, that Hamlet had a heart); nonetheless, they are prescribed to be imagined *should the question arise.* Games of make-believe involve certain explicit or implicit understandings (what Walton calls principles of generation) as to what someone playing the game is to imagine to be true. Walton thinks that some principles of generation "are never explicitly agreed on or even formulated, and imaginers may be unaware of them, at least in the sense of being able to spell them out."[30] But those who play the game correctly have an implicit understanding of what is to be imagined, though in some cases, uncertainty and disagreement may persist.

Walton emphasizes that truth and fictionality are not mutually exclusive. Some of the propositions prescribed to be imagined (by a historical fiction like *War and Peace*, for example) may be true. Even if all the sentences in a novel about the future like George Orwell's *1984* turned out to be true, that novel still remains fiction—similarly in the case of a novel whose sentences accurately describe events on some unknown planet in the universe.[31] What is essential to fictionality is not falsehood but a prescription to be imagined.

Mary Leng is an anti-realist philosopher of mathematics who embraces Walton's theory of fiction in order to deal with the alleged commitments of our best scientific theories to mathematical objects.

28. Kendall L. Walton, *Mimesis as Make-Believe: On the Foundations of the Representational Arts* (Cambridge, MA: Harvard University Press, 1990).
29. Ibid., p. 39. "Fictional propositions are propositions that are *to be* imagined—whether or not they are in fact imagined."
30. Ibid.
31. Ibid., p. 74.

Leng regards the mathematical objects appearing in our scientific theories as merely useful fictions. She explains,

> It is reasonable to view the assumption that there are mathematical objects that are related to non-mathematical objects in various ways as generative of a *fiction*, which provides us with a means, indirectly, to represent non-mathematical objects and their relations. Merely *pretending* that there are mathematical objects satisfying the axioms of a given mathematical theory provides us with a means to represent non-mathematical objects (as being related to these mathematical objects in certain ways). . . . *all* that is confirmed by our empirical successes is that these hypotheses are indeed fictional. . . , that is, that they are correct in their representation of the non-mathematical objects they concern.[32]

In support of the plausibility of her view, Leng provides two non-mathematical examples of the use of fictions in scientific theorizing: **idealizations** (like ideal gases) and **theoretical entities** (like certain fundamental particles). These two cases illustrate the point that even in our best scientific theories a reason to speak *as if* an object existed is not always a reason to believe that that object exists. Leng thinks that nothing in our current, best scientific worldview gives us good reason to believe in mathematical objects as anything more than useful fictions. Hence, we have no reason to think that such objects exist.

4.1.4.2 Assessment

An important objection to pretense theory is that mathematics is not of the genre of fiction. Mathematicians and scientists take mathematics to be a body of knowledge and a realm of discovery, not invention. To regard mathematical statements as fictional is alleged to distort the nature of this discipline.

In weighing this objection, we need to keep in mind that the pretense theorist is not defending a hermeneutical claim about how professional mathematicians or scientists in fact understand mathematical sentences. In order to undercut the Indispensability

32. Mary Leng, *Mathematics and Reality* (Oxford University Press, 2010), pp. 9–10.

Argument for Platonism, the pretense theorist need show only that mathematical sentences *can* be reasonably taken to be fictional. In fact, there are some features of mathematics that make it seem a prime candidate for a fictional interpretation. For example, **axiomatization** of mathematical theories naturally invites a pretense theoretical interpretation. Take the **Axiom of Infinity** in standard **Zermelo-Fraenkel set theory:**

> Axiom of Infinity: There exists a set x having the empty set $\varnothing$ as a member, and for any member y of x the union of y and $\{y\}$ is also a member of x.

It is striking that this is the only axiom of standard set theory with existential implications. On a pretense theoretical approach, the Axiom of Infinity is something we are prescribed to imagine true. We are to make believe that there is an infinity of these things called sets and then are free to explore the fictional world of our imagination. This will certainly be a journey of discovery, which will issue in a great deal of knowledge of the mathematical world determined by the axioms. Such an attitude toward the axioms of set theory is not uncommon among mathematicians and philosophers of mathematics. For example, **postulationalism**, which treats the axioms of competing theories as postulates whose consequences may be explored, invites us, in effect, to make believe that the axioms are true without committing ourselves to their objective truth.[33] In fact, the philosopher of mathematics Stewart Shapiro observes, "The strongest versions of working realism are no more than claims that mathematics can (or should) be practiced *as if* the subject matter were a realm of independently existing, abstract, eternal entities."[34] This characterization would make working realists into pretense theorists!

But is it plausible to take the Axiom of Infinity as something prescribed to be imagined rather than as a straightforward metaphysical assertion? I think the answer is obviously affirmative. It is universally admitted that the Axiom of Infinity is not intuitively obvious. Its lack of intuitive warrant was one of the heavy stones that helped to sink logicism, an early twentieth-century attempt to

33. See Michael Potter, *Set Theory and its Philosophy* (Oxford: Oxford University Press, 2004), pp. 6–11.
34. Stewart Shapiro, *Philosophy of Mathematics: Structure and Ontology* (Oxford: Oxford University Press, 1997), p. 7.

derive set theory from logic alone. Since the axiom lacks intuitive warrant, the Axiom of Infinity is adopted by contemporary mathematicians for reasons that are variously called "pragmatic" or "regressive" or "extrinsic," reasons which do not justify its truth, but its mathematical utility. Moreover, as Potter points out, there are multiple versions of the Axiom of Infinity which postulate different sets. A defense of the axiom on pragmatic grounds "does not directly give us a ground for preferring one sort of axiom of infinity over another," he says, since any of them will work.[35] As a result, says Potter, Platonists "have frustratingly little they can say" by way of justification for their preferred version of the axiom.[36] As a serious assertion of ontology, the Axiom of Infinity is a breathtaking assertion that utterly outstrips our intuitions.

So much more can be said on this head, but axiomatic set theory seems to be a perfect candidate for a pretense theoretical approach. This conclusion has sweeping significance because set theory is typically regarded as foundational for the rest of mathematics, since the whole of mathematics can be reductively analyzed in terms of pure sets. In the words of Penelope Maddy, "our much-valued mathematical knowledge rests on two supports: inexorable deductive logic, the stuff of proof, and the set theoretic axioms."[37] If the latter involve no ontological commitments, neither does the whole of mathematics.

Summary

The Platonist certainly has his work cut out for him if he is to provide adequate warrant for our affirming Premiss (II) of the Indispensability Argument. Wholly apart from proving the untenability of realist versions of anti-Platonism, he must defeat all of the defeaters offered by anti-realist alternatives to taking mathematical sentences to be literally true. Since the neo-Quinean himself sanctions the use of paraphrases in order to avoid unwanted ontological commitments, he faces the daunting task of showing that neither modal structuralism nor constructibilism successfully offers just such paraphrases.

(*Continued*)

35. Potter, *Set Theory*, p. 70.
36. Ibid., p. 72.
37. Ibid., p. 1; see further Chapter 2, "Set Theory as a Foundation."

(*Continued*)

Moreover, he must show why we may not take mathematical discourse to be a kind of figurative language involving the use of existential metaphors, thereby avoiding commitment to mathematical objects. He must show why, despite common mathematical practice, axiomatic set theory may not be interpreted as proposing postulates which we are to imagine to be true without asserting them to be literally true. On top of it all, after all is said and done, the Platonist needs to explain why we cannot take the metaphysically heavy statements of mathematics, Platonistically construed, to be simply false, however useful they may be in scientific practice and everyday life.

4.2 Responses to Premiss (I)

The foregoing strategies all accept or, at least, do not challenge the neo-Quinean criterion of ontological commitment that lies at the heart of the Indispensability Argument. More fundamental challenges to the Indispensability Argument call into question that criterion and, hence, Premiss (I). Some of these versions of anti-Platonism ask why true sentences may not contain irreferential singular terms or why successful reference must involve the existence of an object in the world. Others may challenge the neo-Quinean criterion by demanding why use of the existential quantifier should be interpreted to assert the existence of an object which is the value of the variable bound by the quantifier.

4.2.1 Arealism

4.2.1.1 Exposition

As we saw, Carnap drew a fundamental distinction between what he called "internal questions" and "external questions," that is to say, questions which are posed *within* an adopted linguistic framework and questions posed by someone *outside* that framework. Carnap gives the illustration of what he calls the "thing" framework or language. Once we have adopted the linguistic framework in which we speak of observable things in a spatiotemporal system, we can meaningfully pose internal questions like "Is the Moon a thing?" or "Is a school of fish a thing?" Someone who rejects the language of things may choose to speak instead of mere appearances such as sense data.

Similarly, once we have adopted a linguistic framework involving terminology for abstract objects like numbers, internal questions like "Is there a prime number greater than 100?" are meaningful. No one who has adopted the framework would seriously raise the question "Are there numbers?," for their existence is necessary once one has adopted the number framework. For someone who is outside the framework, Carnap insisted, the question is meaningless. Quine's criterion of ontological commitment thus applies only within a linguistic framework, the adoption of which is a matter of convention.

4.2.1.2 Assessment

No philosopher today would defend Carnap's verificationism; but his conventionalism does find an echo today in what is sometimes called **ontological pluralism.**[38] According to pluralists, certain ontological questions, though meaningful, do not have objective answers. The philosophers of mathematics Mark Balaguer and Penelope Maddy, for example, would deny that the question "Do mathematical objects exist?" has an answer that is objectively true or false.[39] On arealism, there just is no fact of the matter whether or not mathematical objects exist.

It is extraordinarily difficult, however, to understand how there could be no fact of the matter whether or not abstract objects exist, short of a radical ontological pluralism according to which for some language users, there *really* are abstract objects, whereas for

38. See David J. Chalmers, "Ontological Anti-Realism," in *Metametaphysics: New Essays on the Foundations of Ontology*, ed. David Chalmers, David Manley, and Ryan Wasserman (Oxford: Clarendon, 2009), pp. 77–129. Chalmers' terminology is misleading, since anti-realism on the level of **ontology** involves the denial of the existence of mathematical objects. It is only on the metaontological level that anti-realism is the denial that the ontological question has an objective answer. So it would be less misleading to call ontological pluralism "metaontological anti-realism" rather than "ontological anti-realism."
39. See Balaguer, *Platonism and Anti-Platonism in Mathematics*, pp. 151–79; Penelope Maddy, *Defending the Axioms: On the Philosophical Foundations of Set Theory* (Oxford: Oxford University Press, 2011), p. 98. Maddy offers various characterizations of arealism.

other language users, there are not.[40] Such an ontological relativity might strike us as utterly fantastic (were there numbers in Jones' world during the Jurassic Period because Jones would one day affirm mathematical statements?).

This negative verdict on an arealist solution does not, however, imply that Carnap's analysis is without merit. For despite the widespread rejection of conventionalism, Carnap's distinction between external and internal questions continually re-surfaces in contemporary discussions and strikes many philosophers as intuitive and helpful. By distinguishing between internal and external questions, the anti-realist circumscribes Quine's criterion of ontological commitment. Quantification in the language in which internal questions are posed is not ontologically committing for the person speaking outside the framework. Only quantification in the language in which external questions are posed would commit us to the mind-independent reality of the objects quantified over. Statements in the external language which are analogues of internal statements may avoid quantifying over abstract objects (*ultima facie* strategies) or may be taken to be simply false (fictionalism) or to be neutral in their ontological commitments (neutralism). All of these anti-realisms are, of course, controversial; but what is clear is that the debate has moved far beyond the flat imposition of the neo-Quinean criterion.

4.2.2 Free Logic

4.2.2.1 Exposition

Free Logics are logics which are free of existential import with respect to singular terms but whose quantifiers are taken to be devices of ontological commitment. This involves an essentially Quinean view of ontological commitment: we commit ourselves to the existence of certain things, not by the use of singular terms in sentences we take to be true, but exclusively through the use of first-order quantifiers. Free logic will avoid the unwanted commitments which Quine's criterion would engender by scrapping classical logic's inference rules of **Existential Generalization** (EG) and **Universal**

40. For discussion see Matti Eklund, "Carnap and Ontological Pluralism," in *Metametaphysics*, pp. 137–51; cf. idem, "The Picture of Reality as an Amorphous Lump," in *Contemporary Debates in Metaphysics*, ed. John Hawthorne, Theodore Sider, and Dean Zimmerman (Oxford: Blackwell, 2008), pp. 382–96.

Instantiation (UI). So, for example, from the truth of "Sherlock Holmes is the most famous detective of English fiction" we cannot infer that $\exists x$ (x = the most famous detective of English fiction). By the same token, from the arithmetic truth that $3 < 5$ we cannot infer that $\exists x$ ($x < 5$).

A view that asks us to abandon EG and UI would seem to require very powerful motivations. It is perhaps surprising how powerfully motivated some of the claims of free logic are. Karel Lambert, a pioneer of free logic during the 1950s, complains that although modern logic in the late nineteenth century shed itself of various existence assumptions implicit in Aristotelian logic with respect to the use of general terms,[41] modern logic remains infected with existence assumptions with respect to the use of singular terms, assumptions that ought not to characterize a purely formal discipline.[42] For we have at the deepest level,

> A primordial intuition that logic is a tool of the philosopher and ideally should be neutral with respect to philosophical truth. … So if there are preconditions to logic that have the effect of settling what exists and what does not exist, they ought to be eliminated because they corrupt the ideal of logic as a philosophical tool.[43]

---

41. These include the assumptions that in the traditional "square of opposition" A-statements like "All men are mortal" imply I-statements like "Some man is mortal," and E-statements like "No men are immortal" imply O-statements like "Some man is not immortal." Modern sentential logic strips Aristotelian logic of these existence assumptions by interpreting universally quantified statements to have the logical form of conditionals, e.g., "If something is a man, then it is mortal," which carries no commitment to the existence of a man. Just as modern sentential logic aspires to be free of existence assumptions with respect to general terms, so free logic aspires to go one step further to be free of existence assumptions when it comes to singular terms.
42. Karel Lambert, "Existential Import Revisited," *Notre Dame Journal of Formal Logic* 4 (1963): 288–92; idem, "The Nature of Free Logic," in *Philosophical Applications of Free Logic,* ed. Karel Lambert (Oxford: Oxford University Press, 1991), pp. 3–12; idem, *Free Logic* (Cambridge: Cambridge University Press, 2003), pp. 17–24.
43. Karel Lambert, *Meinong and the Principle of Independence* (Cambridge: Cambridge University Press, 1983), pp. 98–9. For example, shouldn't the methods of logic apply to reasoning involving terms which, for all we know, may or may not refer to existing objects, as in the case of astronomers who used "Vulcan" before knowing whether such a planet existed or not? John Nolt observes that the obligation to confirm the existence of things before naming

These existence assumptions regarding singular terms surface dramatically in the way in which standard modern logic handles identity statements. For such statements cannot be true, according to standard logic, unless the referents of the singular terms employed in such statements exist. In other words, identity statements are ontologically committing for him who asserts them. But it seems bizarre to think that from a seemingly tautologous truth of the form $t = t$, where t is some singular term, it follows that the thing denoted by t actually exists. Nevertheless, this is what standard logic requires. Lambert takes this ontological implication of mere identity statements to be absurd. For it would follow from the fact that "Vulcan = Vulcan" that there is some object identical with Vulcan, that is to say, that Vulcan exists! Standard logic avoids this untoward result by restricting the terms in true identity statements to those designating existing objects. For example, standard logic must regard a statement like "Vulcan = Vulcan" as false, even though it appears to be a tautology which is necessarily true. Standard logic cannot therefore distinguish the truth value of identity statements like "Zeus = Zeus" and "Zeus = Allah." Yet the first seems necessarily true and the second obviously false. Nor can standard logic affirm the truth of "Aristotle = Aristotle," since Aristotle no longer exists and so there is no thing with which he can be identified. It would be the height of ontological presumption, I think, to claim that the truth of such a statement implies a **tenseless theory of time**, according to which all moments and things in time are equally existent. Such an inference would only underscore the free logician's claim that modern logic is still infected with inappropriate existence assumptions. As a result of limiting truths of identity to those whose singular terms denote existing objects, standard logic becomes limited in its application to certain inferences and does not permit us to discriminate between inferences where the referentiality of the terms is crucial and those where it is not. For example, we are prohibited from inferring, "Lincoln was the Great Emancipator; Lincoln brooded; therefore, the Great Emancipator brooded," an inference whose obvious validity should not be dependent on Lincoln's existing.

---

them is so irksome that "even mathematicians routinely flout it. . . . They get away with this, usually, only by being discreetly inexplicit about the underlying logic—which is, in consequence, not rigorously classical" (John Nolt, "Free Logics," in *Handbook of the Philosophy of Science*: vol. 5: *Philosophy of Logic*, ed. Dale Jacquette [Amsterdam: Elsevier, 2006], p. 1023).

Proponents of free logic therefore propose to rid logic of all existence assumptions with respect to both general and singular terms. Free logic has thus become almost synonymous with the logic of **irreferential** (or non-denoting, vacuous, empty) **singular terms.** Thus, unlike neo-Meinongianism (to be discussed below), free logic need not presuppose that the referents of such terms are non-existent objects; rather, there just are no referents. Advocates of so-called **positive** (as opposed to negative or neutral) **free logic** maintain that certain sentences can be truly asserted even though they contain irreferential singular terms.

4.2.2.2 Assessment

This feature of positive free logic strikes me as well-motivated and eminently plausible. The truth of identity statements involving vacuous singular terms is of a piece with the assumed truth of many sentences which feature vacuous singular terms. Michael Dummett cites as an illustration the following paragraph from a London daily:

> Margaret Thatcher yesterday gave her starkest warning yet about the dangers of global warming caused by air pollution. But she did not announce any new policy to combat climate change and sea level rises, apart from a qualified commitment that Britain would stabilize its emissions of carbon dioxide—the most important 'greenhouse' gas altering the climate—by the year 2005. Britain would only fulfill that commitment if other, unspecified nations promised similar restraint.[44]

Dummett then observes, "Save for 'Margaret Thatcher,' 'air' and 'sea,' there is not a noun or noun phrase in this paragraph incontrovertibly standing for or applying to a concrete object (is a nation a concrete object, or a gas?)."[45] Obviously not; but then is a nation or a gas an abstract object? No. Taking the singular terms of this paragraph to be ontologically committing would commit its user to such strange entities as commitments, dangers, and the climate. If we are not ingenious enough to find acceptable paraphrases for

44. Michael Dummett, *Frege: Philosophy of Mathematics* (Cambridge, MA: Harvard University Press, 1991), p. 231.
45. Ibid.

such sentences, are we really committed to such bizarre entities, on pain of the falsehood of our discourse?

Free logic's denial that use of singular terms in sentences we take to be true is ontologically committing for their user is, I think, quite plausible and constitutes a step in the right direction. But because free logic takes the existential quantifier of first-order logic to carry existence commitments, it cannot avoid the Platonistic commitments of much mathematical talk. Instead, one will have to have recourse to some other anti-realism like fictionalism in order to avoid such unwelcome commitments—unless, that is, one interprets first-order quantifiers to also be ontologically neutral. Such a neutral logic will not, technically speaking, be a free logic, but as Alex Orenstein remarks, such terminology may be misleading, for "Isn't a logic which disassociates the quantifiers from existence a paradigm of a logic that is free of existence assumptions, indeed freer of existence assumptions than Lambert's variety?"[46] The viability of these solutions will occupy us in the sequel.

4.2.3 Neo-Meinongianism

4.2.3.1 Exposition

Are there things that do not exist? Notoriously, the Austrian philosopher Alexius Meinong thought so, and there has been among contemporary analytic philosophers in recent decades a remarkable resurgence of Meinongian thinking with respect to non-existent objects. Since neo-Meinongians typically include abstract objects among the things that do not exist, **neo-Meinongianism** is a congenial option for the anti-Platonist. With respect to the customary devices of ontological commitment, neo-Meinongianism holds that that successful reference does, indeed, involve a relation to an object of some sort but denies that quantification over various objects is ontologically committing.

4.2.3.1.1 SINGULAR TERMS

With respect to singular terms, Richard Routley, the dean of contemporary neo-Meinongians, inveighs against what he calls "the

46. Alex Orenstein, "Is Existence What Existential Quantification Expresses?" in *Perspectives on Quine*, ed. Robert B. Barrett and Roger F. Gibson (Oxford: Basil Blackwell, 1990), p. 265.

Reference Theory" as "the fundamental philosophical error" of the customary semantics.[47] That theory states,

RT. All (primary) truth-valued discourse is referential,

where "referential" is understood to mean that the subject terms of that discourse have as their referents existing objects.[48]

In Routley's view, the sense of "refer" in ordinary English is ontologically neutral, but the word has become **existentially loaded** in philosophical usage.

> The word 'refer' is used in everyday English (see *OED*), in the relevant sense, to indicate merely the subject or topic of discourse, or subject-matter, or even more loosely what such discourse touched upon or what was drawn attention to or mentioned. *Any* subject of discourse can count as referred to, including nonentities of diverse kinds; in this sense there is not commitment to existence. Superimposed on this non-theoretical usage we have a philosophers' usage which embodies theoretical assumptions about language, according to which the reference of a subject expression is some existing item (or extensionally characterised entity) in the actual world.[49]

Because the word "refer" has become so loaded with existence assumptions among philosophers, Routley thinks that we have little choice but to reserve "refer" for the more restrictive relation. So he concedes that a term "*a*" has reference only in the case that *a* exists. Nevertheless, "*a*" can still be "about" or "signify" or "designate" *a*, even though *a* does not exist. Reference can be defined in terms of aboutness: "*a*" refers to $b =_{def.}$ "*a*" is about *b*, and *b* exists. "Then 'refers to' is existentially loaded and can do the classical work of

47. Richard Routley [Sylvan], *Exploring Meinong's Jungle and Beyond: An Investigation of Noneism and the Theory of Items* (Canberra: Research School of Social Sciences, Australian National University, 1979), p. i.
48. Ibid., p. 52. By "primary" Routley means to exclude, I presume, complex sentences involving intensional contexts, such as "Johnny fears the bogeyman," which are not taken to require referents for singular terms within such contexts in order to be true. By "truth-valued" Routley must mean "true," since on the customary semantics a sentence containing a vacuous singular term can have the truth value "false."
49. Ibid., p. 53.

'hooking language onto the world.'"[50] We can, however, use "*a*" as the proper subject of a true statement without using it to refer. For that reason, Routley rejects the Reference Theory (RT).

Although Routley recognizes that much of our discourse is referential, nonetheless RT cannot account for all the data. "There are many true statements of natural language whose truth cannot be reconciled with the RT and the standard ways of attempting to reconcile them with the RT involve unacceptable distortion."[51] Among these are statements about non-entities. Thus, Routley regards as "a major thesis" of neo-Meinongianism the claim that there are true statements in which singular terms occur non-referentially.[52]

Despite Routley's repudiation of RT, he, along with other neo-Meinongians, remains in a sense deeply committed to referentialism with respect to singular terms. For **reference** is taken to be a relation between words and objects, rather than an intentional property of referring agents. Indeed, neo-Meinongianism is a kind of hyper-referentialism in that it takes there to be objects correlated with *every* singular term, whether occurring in a true sentence or not. In this, neo-Meinongians remain faithful to Meinong's own view. Routley concedes that for Meinong every term designates an object, but Routley insists that on Meinong's view not every such object exists.[53] Granted; but Routley is able to deny that such terms successfully refer to their correlative objects only by taking reference to be existentially loaded. Inasmuch as there are, indeed, such correlative objects, no singular terms are truly vacuous or empty, even if they are, in Routley's peculiar, existentially loaded sense, irreferential. Neo-Meinongianism accepts wholeheartedly the assumption of RT that successful reference is a relation in which words stand to certain objects and goes beyond RT in holding that no singular terms fail to stand in such a relation. What it denies is that all of the objects which are the referents of singular terms are existing objects; some, rather, are non-existing objects. For Meinong and his followers, singular terms in true sentences do refer to objects—they are not irreferential—but the objects they refer to may be non-existent.

50. Ibid., p. 617.
51. Ibid., pp. 58–9.
52. Ibid., p. 206.
53. Ibid., p. 490.

4.2.3.1.2 EXISTENTIAL QUANTIFICATION

On the other hand, neo-Meinongianism holds that quantification over various objects is not ontologically committing. Terence Parsons, for example, holds that "$\exists x$" is not ontologically committing and that genuine existence claims should be symbolized by means of an existence predicate *E!*.[54] In such a case, Meinong's famous dictum, "There are objects which do not exist" can be formulated $(\exists x)\ (\neg E!x))$, which resolves any appearance of contradiction. By contrast, Routley is willing to affirm that traditional first-order quantificational logic is, as he puts it, "existentially loaded."[55] Routley sees this as a serious shortcoming of classical logic with respect to capturing many truths and valid inferences of natural language, and so he advocates a revision of classical logic to strip it of its existence assumptions.

Like free logicians, Routley contends that the central truths of logic should be prior to and independent of claims of particular metaphysical theories, for, since they are used in deducing consequences from and thereby assessing those theories, they should not depend for their correctness on those very theories.[56] But while Routley applauds free logic's affirmation of our ability to express subject-predicate truths, where the subject term is empty, he faults free logic for retaining the existentially loaded quantifiers of classical logic. Because free logic permits quantification over only what exists, it proves to be "an unsatisfactory halfway house" on the road to an adequate theory, which is a truly neutral quantification logic.[57] One may retain entirely the formalism of classical logic, including UI and EG, by removing any ontological commitments in pure logic. To signal the difference in interpretation, Routley substitutes the **particular quantifier** "P," to be read "for some. . . ," for the existential quantifier "$\exists$," and the universal quantifier "U" for the traditional "$\forall$". By adding an existence predicate "*E!*" we can

54. Terence Parsons, *Nonexistent Objects* (New Haven, CT: Yale University Press, 1980), pp. 6, 156. See the interesting piece by Graham Priest, "The Closing of the Mind: How the Particular Quantifier Became Existentially Loaded behind our Backs," *Review of Symbolic Logic* 1 (2008): 42–55. He points out that it is really only since Bertrand Russell that, under Quine's influence, the view that "some" expresses existence came to be widely held.
55. Routley, *Exploring Meinong's Jungle*, p. 56.
56. Ibid., p. 75.
57. Ibid., p. 79.

symbolize "Some things do not exist" as (Px) ($\neg E!x$). Routley thus arrives at the same destination as the neutralist, namely, a neutral logic, whose virtues Routley unfolds at length.[58]

4.2.3.1.3 APPLICATION TO MATHEMATICAL STATEMENTS

Meinong himself held enigmatically that mathematical objects have *being* even though they do not *exist*. Neo-Meinongians reject the distinction between being and existence and so regard abstract objects as belonging to the realm of objects which do not have being, since they do not exist. Classical mathematics can be retained without reservation once a neutral quantificational logic is in place. Routley explains that once we give up the assumption that all statements about non-existing items are false,

> We can foresake [*sic*] the easy platonism that even nominalists sometimes slip into over mathematics; for we have nothing to lose (in the way of discourse) by taking a hard, commonsense line on what exists, e.g. that to exist is to be, and be locatable now, in the actual world. We are no longer forced to distinguish being or existence from actuality or to extend 'exists' beyond this sense, e.g. to numbers and to the ideal items of theoretical sciences, simply in order to cope with the fact that apparently nonexistent items figure frequently in many calculations and in much theory: for we may retain the (perhaps redrafted) theory while admitting that the items do not exist.[59]

In the now re-drafted theory, being and existence are the same, and abstract objects have no being, period. Accordingly, neo-Meinongians think that there are objects, including mathematical objects, that do not exist and, moreover, that we frequently refer to such objects in literally true sentences.

The neo-Meinongian, then, would undercut the Indispensability Argument for Platonism by rejecting the customary criterion of ontological commitment. Quantificational logic is taken to be neutral with respect to ontological commitment. Although abstract

58. Ibid., Chapter 1, part IV, §§15–21.
59. Ibid., p. 45; *cf.* pp. 223–24. *N.B.* Routley's commitment to presentism. Some of his most persuasive examples of truths about non-existent objects will be past- or future-tense statements about individuals who no longer or do not yet exist.

singular terms are referential, their referents do not exist. Therefore mathematical discourse can be true without committing its user ontologically to existent objects.

4.2.3.2 Assessment

What might be said by way of assessment of neo-Meinongianism?

4.2.3.3 Singular Terms

Consider first its treatment of singular terms. Since Routley takes "refer" to be existentially loaded, he repudiates RT, which requires that singular terms appearing in extensional contexts in true sentences are referential, that is to say, have existing referents. As we have seen, however, Routley still assumes reference to be a relation between words and objects rather than an activity of persons. Indeed, on his account all singular terms, whether in true sentences or not, have objects picked out by them. There are, on his view, no truly vacuous or irreferential singular terms in the sense of terms to which there are no correlated objects. Singular terms may be irreferential only in the sense that the objects to which they are related do not exist. So what Routley objects to in RT is not that reference is construed as a relation between words and objects but rather that the objects which stand in relation to certain words must be existing objects.

It seems to me, therefore, that Routley's critique of RT is not as radical (in the sense of getting to the root of the matter) as it ought to be. J. N. Findlay indicts Meinong's Theory of Objects primarily on these grounds. He writes,

> Meinong assumes throughout his treatment that an object is in some sense a logical *prius* of a conscious reference or intention: for there to be a conscious reference or intention there must in some wide sense *be* something which that reference or intention is 'of'. . . . Since this is obviously not true in an ordinary sense in the case of *some* conscious intentions, there must, Meinong thinks, be a subtle sense or senses in which there *are* objects of such conscious intentions. . . . What should be seen is that this whole line of argumentation is wrong: . . . what our usage shows is that 'thinking' and its cognates are not relational expressions like 'above', 'before', 'killing', 'meeting',

> &c., nor can they be said to express relations. . . . We cannot therefore validly take an object of thought out of its object-position in a statement and make it an independent subject of reference: from '*X* thinks of a *Y* as being *Z*' it does not follow that there is a *Y* which is being thought of by *X*, nor even that a thought-of *Y* really is *Z*. . . .
>
> That intentionality is not a relation but 'relation-like' (*relativliches*) is, of course, an insight of Brentano's: Meinong, who frequently surpassed his master, in this respect certainly lagged behind him. He could only conceive intentional references in terms of objects **logically prior** to them, on which they necessarily depended: hence, the many absurdities of his theory of objects.[60]

Meinong's Theory of Objects should be accompanied by a more radical critique of RT, so as to call into question the assumption that reference is a word-object relation. Routley does not, to my knowledge, directly engage the Brentano-Husserl-inspired construal of reference as an intentional property of agents, nor have I encountered such an engagement in other neo-Meinongian thinkers.[61] Launching such a fundamental critique of RT would, however, be to abandon neo-Meinongianism for neutralism, the next and final option we shall consider.

4.2.3.3.1 EXISTENTIAL QUANTIFICATION

With respect to existential quantification, it seems to me that neo-Meinongians have argued persuasively that a neutral quantificational logic is preferable to a logic with existentially loaded quantifiers. Consider four areas where the use of existentially loaded quantifiers results in an inadequate treatment of important philosophical concerns.

60. J. N. Findlay, *Meinong's Theory of Objects and Values*, 2nd ed. (Oxford: Clarendon Press, 1963), pp. 343–44.
61. But see the very nice discussion of Brentano's views on **intentional** objects in Kenneth J. Perszyk, *Nonexistent Objects: Meinong and Contemporary Philosophy*, Nijhoff International Philosophy Series 49 (Dordrecht: Kluwer Academic Publishers, 1993), pp. 198–202.

4.2.3.3.1.1 Intentional Statements

First, we shall be logically hamstrung when it comes to dealing with intentional statements. It will be recalled that the criterion of ontological commitment proposed in (I) does not apply when it comes to intensional **contexts.**[62] But statements of human intentions set up intensional contexts. For example:

1. Ponce de Leon was searching for the Fountain of Youth.

Taking our quantifiers to be existentially loaded, we cannot quantify into such a context, so as to infer "There was something that Ponce de Leon was searching for," since that would imply that the Fountain of Youth exists! Routley is emphatic about there being true intentional statements in which things that do not exist are referred to or quantified over. He observes, "The overwhelming part of everyday, and also of extraordinary, of scientific and of technical discourse is intensional."[63] Intentional statements are universally regarded as true. But they are not generally regarded as genuine statements about things that do not exist because intensional contexts are "opaque," that is to say, one cannot quantify into such contexts or refer to the objects which are the denotations of the singular terms appearing in those contexts.

But why not? This exemption is simply the unpleasant consequence of adopting the customary criterion of ontological commitment, if that criterion is to be at all plausible. Exempting intensional frameworks might well be seen as an ad hoc adjustment aimed at saving an otherwise implausible criterion. On a neutral logic, quantifying into intensional contexts is not ontologically committing and so need not result in ontological extravagances. So from (1) above, one may validly infer.

2. There is something that Ponce de Leon was searching for,

meaning merely that the explorer was not aimlessly rambling about but had an object in mind, for which he was looking. This is not to say that in certain situations the assertion of a statement like (2) does not convey and would not be taken to convey the speaker's

62. See p. 86.
63. Routley, *Exploring Meinong's Jungle*, p. 8. He presents his theory of o the key to **intentionality.**

conviction about the real existence of something. It is merely to insist that quantificational phrases are not in virtue of their meaning ontologically committing. Whether a speaker is ontologically committed to the thing quantified over will vary from situation to situation.

Peter van Inwagen exploits the situational relativity of neutral quantificational phrases to try to make the neo-Meinongian position look silly:

> One day my friend Wyman told me that there was a passage on page 253 of volume IV of Meinong's *Collected Works* in which Meinong admitted that his theory of objects was inconsistent. Four hours later, after considerable fruitless searching, I stamped into Wyman's study and informed him with some heat that there was no such passage. 'Ah,' said Wyman, 'you're wrong. There is such a passage. After all, you were looking for it: there is something you were looking for. I think I can explain your error; although there *is* such a passage, it doesn't *exist*. Your error lay in your failure to appreciate this distinction.' I was indignant.[64]

In this admittedly funny story, Wyman, presumably an adherent of neutral logic, is prepared to quantify into an intensional context without ontological commitment: after all, van Inwagen was not just browsing through Meinong; there was something he was intent on finding. By contrast, van Inwagen, a Quinean, knows that given his own criterion of ontological commitment, quantifying into such a context would lead to an ontological falsehood, so van Inwagen exempts intensional contexts from the quantifier's reach.

But with what justification? The implication of the joke is that neutral logic would lead us into egregious misunderstandings. But that is surely not the case. Any neutralist would recognize that Wyman's original assertion would be taken to imply that upon turning to page 253 of volume IV one might read Meinong's confession and that van Inwagen's indignant protest that there is no such

64. Peter van Inwagen, *Ontology, Identity, and Modality: Essays in Metaphysics*, Cambridge Studies in Philosophy (Cambridge: Cambridge University Press, 2001), p. 16. As the name "Wyman" reveals, what we have here represented is the caricature of Meinongianism given by Quine in his influential essay "On What There Is," *Review of Metaphysics* 2 (1948): 21–38, not genuine Meinongianism. See comments by Graham Priest, *Towards Non-Being: The Logic and Metaphysics of Intentionality* (Oxford: Clarendon Press, 2005), p. 108.

passage was intended to convey the opposite. Everyone knows that "there is" is often used in ordinary language to express the existence of real world objects—but, according to the neutralist, not always and automatically. There are contextual clues and rhetorical devices by means of which we can make our intentions clear. Ironically, in his discussion of neo-Meinongianism van Inwagen himself uses the device of italics to signal an ontologically relevant use of "there is": "There *is* no nonexistent poison in the paranoid's drink. There *is* no such thing as his uncle's malice."[65] Thus, by double italicizing Wyman's statement "there *is* such a passage, but it doesn't *exist*," van Inwagen fosters the desired appearance of contradiction.

Similarly, consider van Inwagen's handling of A. J. Ayer's claim that

> it makes sense to say, in a case where someone is believing, that there is something that he believes. But it does not follow from this that … something must exist to be believed, in the way that something must exist to be eaten or to be struck.[66]

Van Inwagen's claim that Ayer's words involve "a glaring apparent contradiction" is purchased only by imposing on Ayer the neo-Quinean's uniformly existentially loaded quantifiers. But Ayer's first sentence is a neutral quantificational claim; the second is an existentially loaded claim. You get a contradiction only by assuming that the first order existential quantifier inevitably carries ontological commitments, which is exactly what neo-Meinongians deny. Only by assuming that Ayer's quantificational claims involve ontological commitment does the neo-Quinean generate a contradiction where there was none.

Van Inwagen is thus assuming his own criterion of ontological commitment when he issues the following challenge to the neo-Meinongian: "If you think there are things that do not exist, give me an example of one. The right response to your example will be either, 'That does too exist,' or 'There is no such thing as

65. Van Inwagen, *Ontology, Identity, and Modality*, p. 16; *cf*. idem, "Fiction and Metaphysics," *Philosophy and Literature* 7:1 (1983): 68.
66. A. J. Ayer, *Thinking and Meaning* (London: H. K. Lewis, 1947). For van Inwagen's remarks, see Peter van Inwagen, "Response to William Lane Craig's *God over All*," *Philosophia Christi* 21 (2019): 267–75.

that'."[67] Here the two prongs of the "right response" make it evident that van Inwagen is assuming that "there is" carries ontological commitment, so that the second prong "There is no such thing as that" means to deny that the thing in question does exist. So if I, in response to his challenge, say, "the hole in my shirt," a retort by van Inwagen, "That does, too, exist" seems ontologically extravagant; while the retort, "There is no such thing as that" means only that the hole in my shirt does not exist in a metaphysically heavy sense, which is precisely what I already affirmed! On a neutral logic, I can quantify over holes in my shirt without ontological commitment to holes as existing things. To return to the subject of intensional contexts, given a neutral logic, I can quantify into an intensional context like "Peter contemplated the hole in my shirt" by asserting "There is a hole in my shirt which Peter contemplated" without ontological commitment to holes, whereas an existentially loaded logic will be hobbled in its permitted inferences by the presupposition of the neo-Quinean criterion of ontological commitment.

4.2.3.3.1.2 Tensed Statements

A second respect in which existentially loaded quantification is markedly disadvantageous concerns tensed statements about no longer existent or not yet existent individuals. Routley is a **presentist**, who denies that the past and future are on an ontological par with the present. I have elsewhere argued at length in favor of such a **tensed theory of time.**[68] Even if some partisans of tensed time deny that future contingent statements are true or false, nonetheless, very few have the temerity to deny that past-tense statements about individuals who do not presently exist are true. For example, it seems indisputably true that "There have been 45 U.S. presidents." The non-existence of most of them is no impediment to our quantifying

67. Van Inwagen, *Ontology, Identity, and Modality*, p. 16; similarly, idem, "A Theory of Properties," in *Oxford Studies in Metaphysics*, ed. Dean Zimmerman (Oxford: Oxford University Press, 2004), p. 129; idem, "Being, Existence, and Ontological Commitment," in *Metametaphysics: New Essays on the Foundations of Ontology*, ed. David J. Chalmers, David Manley, and Ryan Wasserman (Oxford: Oxford University Press, 2009), p. 481.
68. See my *The Tensed Theory of Time: A Critical Examination*, Synthèse Library 293 (Dordrecht: Kluwer Academic Publishers, 2000) and my *The Tenseless Theory of Time: A Critical Examination*, Synthèse Library 294 (Dordrecht: Kluwer Academic Publishers, 2000).

over past U.S. presidents. To infer from the truth of such statements that time is, in fact, **tenseless** and that past and future individuals are on an ontological par with present individuals would be to draw a breathtaking metaphysical inference on the basis of the slim reed of the neo-Quinean criterion of ontological commitment.

It is noteworthy that in debates over **presentism**, tenseless time theorists tend simply to presuppose without argument that quantification is ontologically committing, and so our ability to quantify over past/future individuals in true sentences is taken to commit us to their existence.[69] Thus, Theodore Sider muses, "Since ordinary talk and thought are full of quantification over non-present objects, presentists are in a familiar predicament: in their unreflective moments they apparently commit themselves to far more than their ontological scruples allow."[70] In fact, the only person in a predicament here is the presentist who also accepts the neo-Quinean criterion of ontological commitment. It never seems to occur to tenseless time theorists that our ability to quantify over purely past/future individuals in true sentences might be a good reason to reject the criterion of ontological commitment which they unquestioningly presuppose. Sider himself thinks that the presentist should adopt a sort of fictionalism: sacrifice the truth of past- and future-tense statements in favor of their quasi-truth or verisimilitude. There is no reason the presentist should accept such a fool's bargain, for it is far more obvious that, for example, the statement "Some medieval theologians wrote in Latin" is true than that the neo-Quinean criterion of ontological commitment is true. Sider notwithstanding, we *do* speak as if presentism is true, but we do *not* speak as if quantification is a device of ontological commitment. That is why we quantify freely over merely past objects without thinking that such objects therefore exist.

---

69. See, e.g., David Lewis, "Tensed Quantifiers," *Oxford Studies in Metaphysics* 1 (2004): 3–4. Notoriously, Lewis also presented a similar quantificational justification for belief in modal realism on the grounds that there are ways the world might have been. Not only is such a metaphysical commitment to the reality of other worlds extravagant, but by the same token there are also, then, impossible worlds, ways the world could not have been (Margery Bedford Naylor, "A Note on David Lewis's Realism about Possible Worlds," *Analysis* 46 [1986]: 28–9; Takashi Yagisawa, "Beyond Possible Worlds," *Philosophical Studies* 53 [1988]: 175–204), which reduces modal realism to absurdity.
70. Theodore Sider, "Presentism and Ontological Commitment," *Journal of Philosophy* 96 (1999): 325.

The tensed time theorist can hardly be blamed if he elects to deny the neo-Quinean criterion of ontological commitment rather than surrender his view of time. Of course, if we had strong independent grounds for preferring a tenseless view of time, then Routley's argument would be void. But we have, in my view, no such grounds.[71]

The case of tensed truths serves as just another illustration of Routley's first point about logic's inability to handle intensional contexts when the quantifiers are taken to be existentially loaded. From "Aristotle was born at Stagira," for example, we cannot infer that "Someone was born at Stagira," since that would involve quantification over a non-existent. The advantage of neutral logic is that it allows us to quantify into such intensional contexts without the need of elaborate paraphrase or committing ourselves to the existence of the things quantified over.

4.2.3.3.1.3 Modal Statements

A third and related respect in which existentially loaded quantification threatens to lead to metaphysically outrageous conclusions concerns **modal discourse.** Oxford logician Timothy Williamson has argued on purely logical grounds for the radical thesis that everything that exists exists necessarily.[72] Williamson calls this view **necessitism. Contingentism** is just the denial of this view, namely, that some things exist contingently.

In order to accommodate the apparent fact that many concrete objects do not exist necessarily, Williamson maintains that such objects could have been and, indeed, once were or will be, abstract objects (or, at least, non-concrete objects).[73] So, for example, I, as a necessary being, could have been and once was a non-concrete object. I existed prior to my conception, not, of course, as a concrete,

71. See further Trenton Merricks, *Truth and Ontology* (Oxford: Clarendon Press, 2007), who challenges the argument that past- and future-tense truths require the tenseless existence of truthmakers of such statements.
72. Timothy Williamson, *Modal Logic as Metaphysics* (Oxford: Oxford University Press, 2013).
73. Williamson recoils from calling objects which are contingently non-concrete "abstract" because they do not fit in with paradigmatic abstract objects like mathematical objects (Ibid., p. 7). But abstract objects come in a bewildering variety, and by the definition of "abstract" as "causally impotent" such non-concrete objects qualify as abstract objects. So it is not unfair, I think, to say that on Williamson's view I could have been and, indeed, once was, an abstract object.

spatiotemporal object, but as a non-concrete object which later became concrete. What sort of object was I? I was once a merely possible person, an impersonal thing which could be and did become a concrete person, and in worlds in which I never exist as a person I remain a merely possible person.[74] So on Williamson's view I am not essentially a person.

Most people (including professional metaphysicians!) find necessitism to be an outrageously implausible thesis which we should, if at all possible, avoid. It is therefore noteworthy that Williamson's arguments for necessitism depend crucially on taking the quantifiers of first-order logic to be devices of ontological commitment. For example, he compares favorably arguments for necessitism with Lewis and Sider's arguments against presentism based upon quantification over past realities:

> David Lewis challenges presentists to analyse in their terms sentences such as 'There have been infinitely many kings named John' (compare 'There are infinitely many possible stars'), in order to cast doubt on the truth of presentism. Ted Sider challenges sceptics who claim that the dispute between presentism and eternalism is merely verbal to analyse in presentist terms sentences of the form 'Half the objects from all of time that are Ks are Ls' (compare 'There are at least as many possible planets as possible stars').[75]

These lame arguments for a tenseless theory of time, depending as they do on the assumption that existential quantification is ontologically committing, ought, in my opinion, to cast more doubt on the neo-Quinean criterion of ontological commitment than on presentism.

As the parenthetical comparisons in the above citation suggest, Williamson considers the non-paraphraseability of certain quantified

74. Williamson rejects the predicative reading of "possible person," according to which a possible person is a kind of person, in favor of the attributive reading, according to which a possible person is something that could have been a person (Ibid., pp. 10–12). *Cf.* Williamson's correction of A. N. Prior for referring to a merely possible person as "someone" rather than "something" (Ibid., p. 68).

75. Ibid., p. 368; *cf.* p. 150. The references are to David Lewis, "Tensed Quantifiers," *Oxford Studies in Metaphysics* 1 (2004): 3–14; Theodore Sider, "Quantifiers and Temporal Ontology," *Mind* 115 (2006): 75–97.

modal truths like "There are uncountably many possible stars" to indicate ontological commitments on the part of the person who asserts them. The same assumption underlies his boast that necessitists (unlike contingentists) can straightforwardly use quantification over possible worlds.[76]

Williamson makes no attempt at all to justify the assumption that first-order quantification is existentially loaded. He merely explains,

> By definition, ontology concerns what there is. Claims of the form 'There is a G' are naturally paraphrased by first-order sentences of the form $\exists x\ Gx$. The quantification in 'There is a G' is not into predicate position, even if 'G' is replaced by 'property', 'relation', or 'concept'. Ontology is part of metaphysics. The content of the ontological commitment is true if there is a G and false otherwise. Its truth value depends on how the mostly non-linguistic world is, as characterized in first-order terms.[77]

The problem is that sentences of the form "There is a G" or even "G exists" have a very light sense in ordinary English, and so some justification is needed for taking the formal quantifiers that abbreviate them as involving metaphysically heavy commitments.

Williamson's deductive formulations of his arguments for necessitism make it especially clear that he takes first-order quantifiers to carry ontological commitments,[78] for the arguments will typically involve some premise in which it is asserted that $(\exists x)\ (x = y)$, leading to a necessitist conclusion like

NNE. $\Box\ ((\forall x)\ (\Box\ (\exists x)\ (x = y)))$.

That is to say, necessarily, for any x, necessarily there is an object to which x is identical. But on a neutral logic, NNE is itself innocuous. Of course, necessarily, everything is necessarily identical to something, namely, itself; but that does not imply that that thing exists. So, for example, Allah is identical to Allah, but Allah is not identical to Zeus, whether or not either one exists.

76. Williamson, *Modal Logic as Metaphysics*, p. 376.
77. Ibid., p. 261.
78. See ibid., pp. 288–96.

Necessitism illustrates the sort of metaphysical debacle which the neo-Quinean criterion of ontological commitment helps to generate.[79] Williamson declares, "what we want is not a weak modal logic, neutral on the relevant metaphysical questions. Rather, we want a strong modal logic that answers those metaphysical questions."[80] I should think that precisely the opposite is true. We want a **modal logic** that is neutral as to ontological commitments. Williamson complains in another context that "Metaphysics based on weak logic wastes its time taking crank theories seriously."[81] To the contrary, basing metaphysics on an existentially loaded, metaphysically strong modal logic helps to generate a theory like necessitism, which is about as cranky as they get.

4.2.3.3.1.4 Mereological Statements

Finally, a fourth area where the disadvantages of existentially loaded quantification become evident is **mereology**, or the study of parts and wholes. Contemporary debates in mereology have focused upon the so-called **special composition question**, namely, what conditions have to be met in order for a plurality to compose an object.[82] The predominant answer to that question among contemporary metaphysicians is **mereological universalism**. This view assumes a **principle of unrestricted mereological composition** (UMC), according to which any plurality whatsoever compose an object:

> UMC. Necessarily, whenever there are some things, then there is a fusion of those things.

So Lewis, on the basis of UMC, holds that there is such a thing as a trout-turkey, a fusion, for example, of the front half of a

---

79. See Takashi Yagisawa, critical notice of *Modal Logic as Metaphysics*, by Timothy Williamson, *Notre Dame Philosophical Reviews* (October 15, 2013), http://ndpr.nd.edu/news/43612-modal-logic-as-metaphysics.
80. Williamson, *Modal Logic as Metaphysics*, p. 429. Boldly asserting that "Logic has no metaphysically neutral core," Williamson affirms that "on the standard objectual interpretation of the first-level quantifiers, the semantics cannot 'serve as a neutral device for exploring alternative views about possible objects'" (Ibid., p. 146).
81. Ibid., p. 226.
82. For what follows I am indebted to Ross Inman, "On Christian Theism and Unrestricted Composition," paper presented at the annual meeting of the Evangelical Philosophical Society, Milwaukee, WI, November, 2012.

particular trout and the back half of a particular turkey.[83] Given UMC, there are even objects like the sum of me and $\sqrt{2}$. "Though the view presents a wildly counterintuitive description of reality," Ross Inman observes, "there are surprisingly powerful arguments in its favor, which partly explains its widespread appeal among many contemporary metaphysicians."[84] That appeal is so strong that Hud Hudson can write, "the view becoming (if not already) the orthodoxy among those writing in mereological metaphysics is that composition is absolutely unrestricted; any plurality whatever has a mereological sum or fusion."[85]

Does UMC force us to include such wildly counter-intuitive objects as trout-turkeys in our ontological inventory? No; it is only the conjunction of UMC with the neo-Quinean criterion of ontological commitment that would foist such commitments upon us. Hence, some theorists have sought to soften the impact of UMC by denying that it involves ontological commitment to any object in addition to the plurality of objects summed together. Lewis gives the following assurance:

> To be sure, if we accept mereology, we are committed to the existence of all manner of mereological fusions. But given a prior commitment to cats, say, a commitment to cat-fusions is not *a further* commitment. The fusion is nothing over and above the cats that compose it. It just *is* them. They just *are* it. . . . In general, if you are already committed to some things, you incur no further commitment when you affirm the existence of their fusion. The new commitment is redundant, given the old one.[86]

Lewis' backpedaling is eloquent testimony to the implausibility of so bloated an ontology as that apparently sanctioned by UMC. The difficulty for Lewis is that he accepts the neo-Quinean criterion of ontological commitment, so that his metaontology commits him, in spite of himself, to the existence of objects over and above

83. David Lewis, *Parts of Classes*, with an Appendix by John P. Burgess, A. P. Hazen, and David Lewis (Oxford: Basil Blackwell, 1991), p. 80.
84. Inman, "Christian Theism and Unrestricted Composition."
85. Hud Hudson, "Confining Composition," *Journal of Philosophy* 103 (2006): 633.
86. Lewis, *Parts of Classes*, pp. 81–82.

the plurality of objects composing the fusion. He says, "Only if you speak with your quantifiers wide open must you affirm the trout-turkey's existence. If, like most of us all the time and all of us most of the time, you quantify subject to restrictions, then you can leave it out. You can declare that there just does not exist any such thing—*except*, of course, among the things you're ignoring."[87] But, Lewis concedes, "Once you've said 'there is' your game is up."[88]

Hardier advocates of (UMC) do not shrink from admitting the additional objects into their ontology. Thus, an obviously discomfited Lynne Rudder Baker writes,

> So, are sums objects or not? . . . Since I have endorsed universalism, I must (hold my nose and) say that sums are objects. However, the unpalatability of commitment to arbitrary sums is mitigated by the fact that the ontological difference that sums make is negligible: the only ontological effect of holding that sums are objects is to increase the number of existing objects.[89]

We can, however, avoid the noxious smell of mereological universalism simply by abandoning the neo-Quinean criterion of ontological commitment. Given neutral quantifiers, we can talk freely about mereological sums without ontological commitment. Existentially loaded quantification is a positive impediment to mereology and, hence, to metaphysics, an assumption which the discipline would be well without.

The foregoing considerations are sufficient, in my opinion, to warrant rejection of the neo-Quinean view of formal and informal quantifiers. But that conclusion may remain moot: What is important for our purposes, rather, is that these considerations certainly suffice to show that the neo-Quinean interpretation of the existential quantifier is far from incumbent upon us and that therefore we may plausibly reject it. Alex Orenstein wisely reminds us that "There are fashions in philosophical explications," one of which is today taking metaphysically heavy existence assertions to be expressed by the

87. Ibid., p. 80.
88. Ibid., p. 81.
89. Lynne Rudder Baker, *The Metaphysics of Everyday Life: An Essay in Practical Realism* (New York: Cambridge University Press, 2007), p. 193. She says this only after futilely trying to differentiate sums from genuine objects.

first-order existential quantifier.[90] It is perfectly reasonable to buck the fashion trend in this regard. So doing opens the door for affirming truths like "There is a prime number between 2 and 4" without commitment to the existence of mathematical objects.

4.2.4 *Neutralism*

4.2.4.1 Exposition

In our discussion of free logic and neo-Meinongianism, we have already seen reason to question the neo-Quinean view of singular terms and existential quantification as means of ontological commitment. Neutralism, like free logic and neo-Meinongianism, rejects certain key assumptions associated with the neo-Quinean construal of quantification and reference which constitute the common ground shared by fictionalism, *ultima facie* strategies, and pretense theory with Platonism. The neutralist holds that *neither* existential quantification *nor* the use of singular terms is a device of ontological commitment. He is therefore unfazed by the Platonist's conviction that various statements quantifying over or featuring singular terms referring to mathematical objects are, without qualification, true.

4.2.4.2 Assessment

We have already seen in our discussion of neo-Meinongianism and free logic the disadvantages of taking first-order quantifiers and singular terms to be devices of ontological commitment. Here we ask, what reasons might be given for adopting the neo-Quinean approach to ontological commitment?

4.2.4.2.1 EXISTENTIAL QUANTIFICATION

Consider first existential quantification. Jody Azzouni challenges Quine's assumption that the existential quantifier of first-order logic does or should carry the burden of ontological commitment:

90. Alex Orenstein, "Is Existence What Existential Quantification Expresses?" in *Perspectives on Quine*, ed. Robert B. Barrett and Roger F. Gibson (Oxford: Basil Blackwell, 1990), p. 266. Orenstein follows neutralists in introducing an existence predicate instead in order to express existential claims.

> Even if one accepts the idea that scientific theories must be regimented in first-order languages, nothing requires the first-order existential quantifier (despite its logical role in such languages) to carry the burden of *ontological* commitment: one can, that is, regiment a scientific language just the way Quine likes, and simply look elsewhere in the regimented theory for the *ontological* commitments. For example, one can provide a special predicate, 'susceptible to observation' say, or 'causally efficacious,' or, and so on, and recognize the ontological commitments of a discourse to be solely those objects falling under the extension of *that* predicate, to treat only *those* objects as existing (or *real*).[91]

Azzouni complains that Quine never considered having an existence predicate convey ontological commitment rather than the quantifiers.[92]

So why take the quantifiers to play such a decisive role? Quine thought it obvious that the existential quantifier of first order logic is ontologically committing.[93] Azzouni takes Quine's argument for the triviality of the criterion to be that in ordinary language "there is" carries ontological commitment and that this idiom is straightforwardly regimented as the existential quantifier in first order logic. On Quine's view, he says, we must look to ordinary language to see where ontological commitment occurs and replicate it in our regimented language. The existential quantifier naturally reproduces the ontological commitments carried by "there is/are" in the vernacular.

The problem for the Quinean is that in ordinary language informal quantificational phrases do *not* seem to be ontologically committing. Consider Thomas Hofweber's list of some of the things we ordinarily say there are:

- something that we have in common
- infinitely many primes
- something that we both believe
- the common illusion that one is smarter than one's average colleague

91. Jody Azzouni, "On 'On What There Is,'" *Pacific Philosophical Quarterly* 79 (1998): 3.
92. Jody Azzouni, *Deflating Existential Consequence: A Case for Nominalism* (Oxford: Oxford University Press, 2004), p. 52.
93. Quine claimed that his criterion was "scarcely contestable" because "($\exists x$)" is explained by the words "there is an object x such that" (W. V. Quine, *Philosophy of Logic*, 2d ed. [Cambridge, MA: Harvard University Press, 1986], p. 89).

- a way you smile
- a lack of compassion in the world
- the way the world is
- several ways the world might have been
- a faster way to get to Berkeley from Stanford than going through San Jose
- the hope that this dissertation will shed some light on ontology
- the chance that it might not
- a reason why it might not.[94]

It would be fantastic to think that there are real objects answering to these descriptions. There is no evidence that the ordinary language speaker labors under the delusion that there are.[95]

Neo-Quineans may simply prescribe taking "there is/are" and "exists" in such a way as to indicate ontological commitments. When we do, we must correct the ordinary language speaker by either denying the truth of his assertion or offering a paraphrase that avoids the unwanted commitments.

Azzouni takes a very dim view of this sort of move on the part of the neo-Quinean, characterizing it as a kind of philosophical chicanery:

> The literal onticist may also claim one *can* compel the ordinary person into *trying* to paraphrase. 'There are fictional mice that talk,' one says, during a discussion about talking fictional animals. 'Oh, so you believe fictional mice *really exist*?' the philosophical trickster responds. 'I didn't say that,' one responds. 'Yes, you sure did—you said "*there are* fictional mice that talk". What do you think "there is" means?' *One* may now try to paraphrase. But what's happened is that the philosophical trickster has (implicitly) switched on *usage*—now one

94. Thomas Hofweber, "Ontology and Objectivity" (Ph.D. dissertation, Stanford University, 1999), pp. 1–2.
95. We commonly say that there are, for example, shades of gray, differences in height, angles from which something can be seen, principles, hostilities, prospects for success, primes between 2 and 12, hours before dawn, dangerous excesses, drawbacks to the plan, and so on, without imagining that there are mind-independent objects of these sorts. See Gerald Vision, "Reference and the Ghost of Parmenides," in *Non-Existence and Predication*, ed. Rudolf Haller, Grazer Philosophische Studien 25–26 (Amsterdam: Rodopi, 1986), pp. 297–336.

> is speaking so that 'there is' (at least for the time being) *does* convey onticity; and the ordinary person isn't sophisticated enough to see how he or she has been duped.[96]

The trick is co-opting the evident truth of the ordinary language expression in order to imply the truth of a metaphysically heavy-weight assertion. This is illegitimately borrowed capital. As we saw in our discussion of fictionalism, when the alleged ontological commitments of a statement are made plain by the neo-Quinean, it is often not at all obvious that the statement so interpreted is true and its denial false.

Theodore Sider is more candid than the trickster:

> Ontological realism should not claim that ordinary quantifiers carve at the joints, or that disputes using ordinary quantifiers are substantive. All that's important is that one can introduce a fundamental quantifier, which can then be used to pose substantive ontological questions.[97]

When one does so, one is no longer speaking ordinary English, but "a new language—'Ontologese'—whose quantifiers are *stipulated* to carve at the joints."[98] If one stipulates that "there is/are" is being

---

96. Jody Azzouni, "Ontological Commitment in the Vernacular," *Noûs* 41/2 (2007): 224, n. 38. He complains,

> There has been a certain amount of unfortunate condescension towards the ordinary person (and, it should be pointed out, scientists, however sophisticated, are officially among the Quinean laity): a tendency for philosophers to claim that the ordinary person doesn't even know what it *means* to be ontically committed; that is, the discourse uttered by such persons is taken to routinely betray ontic commitments that person would disavow (were they brought to attention). It's a tribute to the rigors of philosophical education that such a view does not appear (to philosophers) implausible *on the face of it*. (Ibid., p. 220)

The stipulationist puts the ordinary person in an awkward situation only by changing the way in which informal quantifiers are used.

97. Theodore Sider, *Writing the Book of the World* (Oxford: Clarendon Press, 2011), p. 171.
98. Ibid., p. 172. Sider says that even if the neutralist "is right about natural language quantifiers, ontologists could always relocate to the metaphysics room, and conduct their dispute in Ontologese" (Ibid., p. 197). See also Peter van Inwagen, "Introduction: Inside and Outside the Ontology Room," in *Existence: Essays in Ontology* (Cambridge: Cambridge University Press, 2014), pp. 1–14. Van Inwagen

used to indicate ontological commitments, then, of course, the neutralist who is an anti-realist will regard a statement like "There are odd numbers" as false, and the neutralist who is a Platonist will regard the statement as true. But neutralism has not thereby been defeated or shown to be irrelevant, for it is a claim about how we *do* commit ourselves ontologically in ordinary English, not in a foreign language like Ontologese. Do we, in fact, commit ourselves ontologically through formal and informal quantifiers, as Quineans have asserted? The answer is plausibly, no.

In fact, the neutralist should protest the suggestion that there even *is* such a language as Ontologese which is spoken in the metaphysics seminar. All there really is is ordinary language, and we can employ various linguistic and rhetorical devices to make it clear when we are speaking in a metaphysically heavy sense. But nobody—not metaphysicians, not ontologists—actually speaks the language of Ontologese, for there just is no such language. Metaphysical discussions would quickly grind to a halt if every participant had to be able to furnish on demand paraphrases of ordinary language sentences involving unwanted ontological commitments when construed as sentences of Ontologese.

Of course, one may simply stipulate that formal and informal quantifiers are to be taken in a metaphysically heavyweight sense. But stipulationists trivialize the debate by simply requiring, in effect, that we must choose between realism and anti-realism about various

uses the metaphor of "being in the ontology room" for speaking Ontologese, or what he calls "Tarskian." Van Inwagen thinks that the sentence "Chairs exist" expresses a different proposition in Tarskian than it does in ordinary language and is therefore false in Tarskian though true in ordinary English. Similarly, I think, if the neutralist were willing to concede that disputants are speaking two different languages, he might say that the proposition expressed by "Numbers exist" in English is true while the proposition expressed by that sentence in Ontologese or Tarskian is false. Van Inwagen actually *agrees* with the neutralist that the proposition asserted outside the ontology room "*is* metaphysically neutral" (Ibid.). If the neo-Quinean demands to know what proposition is expressed by "Numbers exist" in English, the neutralist can say with van Inwagen (Ibid., pp. 15–16) that he has no better way of expressing it than that sentence taken in a lightweight sense. He can illustrate the difference by appealing to examples where paraphrases are available, e.g., the sentence "The number of Martian moons is two" expresses the same proposition in English as "Mars has two moons," which is not ontologically committing to the number 2. But there is no reason to expect that ordinary English should have the resources to paraphrase away all quantification over or reference to things, given the lightweight sense of "exist" in ordinary language.

purported objects. We already knew **that.**[99] What they cannot do is use the obvious truth of ordinary language locutions involving quantificational phrases as justification for accepting the truth of sentences in Ontologese stipulated to carry ontological commitments. The point remains that without a refutation of neutralism, the Indispensability Argument, which is based on how we actually do use formal and informal quantifiers, cannot get off the ground.

4.2.4.2.2 SINGULAR TERMS

But what about neutralism's view of the use of singular terms? It seems to be a datum of ordinary language that we frequently assert true statements which contain singular terms which do not denote existent objects. Consider the following examples:

- The weather in Atlanta will be hot today.
- Sherrie's disappointment with her husband was deep and unassuageable.
- The price of the tickets is ten dollars.
- Wednesday falls between Tuesday and Thursday.
- His sincerity was touching.
- James couldn't pay his mortgage.
- The view of the Jezreel Valley from atop Mt. Carmel was breath-taking.
- Your constant complaining is futile.
- Spassky's forfeiture ended the match.
- He did it for my sake and the children's.

It would be fantastic to think that all of the singular terms featured in these plausibly true sentences have objects in the world corresponding to them. Examples like these are legion. In fact, I suspect that singular terms which refer to real-world objects may actually be the *exception* rather than the rule in ordinary language.

How could it be that we are able to assert truths by means of sentences with empty singular terms? In order to get at this question, we first need to address the question, do vacuous singular terms refer? And in order to answer that question, we need to ask what it is to refer, or what is the nature of **reference**? This question is

99. Unless, that is, we are arealists like Balaguer and Maddy (pp. 106–108).

largely neglected by contemporary theorists. Almost all contemporary theories of reference are actually theories about how to *fix* reference rather than theories about the nature of reference itself. The unspoken assumption behind most contemporary theories of reference is the presupposition that reference is a word-world relation, so that terms which refer must have real world objects as their denotations. It therefore behooves us to look more deeply into the nature of reference.

It is an experiential datum that referring is a speech act carried out by an intentional agent.[100] Words in and of themselves engage in no such activity. Lifeless and inert, words are just ink marks on paper or sounds heard by a percipient. Absent an agent, shapes or noises do not refer to anything at all. If, for example, an earthquake were to send several pebbles rolling down a hillside which randomly came to rest in the configuration JOHN LOVES SUSIE, the names—if we would even call them names–would not refer to anybody. As John Searle says, "Since sentences . . . are, considered in one way, just objects in the world like any other objects, their capacity to represent is not intrinsic but is derived from the Intentionality of the mind."[101] An interpreting agent uses his words as a means of referring to something. Referring is thus an intentional activity of persons, and words are mere instruments.

It is the great merit of Arvid Båve's new deflationary theory of reference that he takes truly seriously the fact, given lip service everywhere, that it is *persons* who refer to things *by means of* their words, so that words at best refer only in a derivative sense, if at all.[102] Båve's significant contribution to our understanding reference is not so much that his theory is **deflationary**, as helpful as that may be, but that he furnishes a central schema for reference formulated in terms of the referring activity of agents.

Båve proffers the following deflationary schema for reference:

(R) *a* refers to *b* iff *a* says something (which is) about *b*,

---

100. A fact emphasized by John R. Searle, *Speech Acts: An Essay in the Philosophy of Language* (Cambridge: Cambridge University Press, 1969), p. 27; *cf.* idem, *The Construction of Social Reality* (New York: Free Press, 1995), p. 228.
101. John Searle, *Intentionality: An essay in the philosophy of mind* (Cambridge: Cambridge University Press, 1983), p. vii.
102. Arvid Båve, "A Deflationary Theory of Reference," *Synthèse* 169 (2009): 51–73. I am very grateful to Arvid Båve for extended discussion of Båve's theory.

where "*a*" always stands for a speaker. Though formulated in terms of agents rather than words, this account is truly deflationary because it does not attempt to tell us anything about the nature of reference itself. It leaves it entirely open whether reference is a relation (as Frege and Meinong assumed) or whether it is an intentional property of a mind (as held by Brentano and Husserl).

Given (R), we now ask, what does it mean to say that *a* says something "about" *b*, as stipulated on the right hand side of the biconditional (R)? Båve proposes to "analyse the expression 'about', and then explain 'refer' in terms of it."[103] He offers the following schema as implicitly defining "about":

(A) That *S* (*t*) is about *t*,

where *S* ( ) is a sentence context with a slot for singular terms. Again, Båve's account of aboutness is extraordinarily deflationary. It does not tell us what aboutness is but simply provides a schema for determining what a that-clause containing a singular term (or, presumably, terms) is about. So, for example, that Ponce de Leon sought the Fountain of Youth is about Ponce de Leon and about the Fountain of Youth because the singular terms "Ponce de Leon" and "the Fountain of Youth" fill the blanks in the sentence context "____ sought____."

Now as a deflationary schema, it is very difficult to see how (A) provides an *analysis* of aboutness or serves as an *explanation* of reference, as Båve claims. Taken as an explanation of reference, (A) seems to get things exactly backwards. The reason why That *S* (*t*) is about *t* is because "*t*" is used by some agent to refer to *t*. It is natural, then, to provide an account of aboutness that is cashed out, not in terms of linguistic expressions, but in terms of the speaker's intentions. What is wanted in place of (A) is something along the lines of

(A′) *a* says something *S* about *b* iff in saying *S a* intends *b*,

where "*a* intends *b*" means something like "*a* has *b* in mind." In accordance with (R) I refer to *b* because I say something about *b*, and in line with (A′) I say something about *b* because in making my

103. Ibid., p. 63.

utterance I intend *b*. On such an account what some theorists have called "**speaker's reference**" becomes paramount.[104]

By construing reference in terms of the referring activity of agents and characterizing aboutness in intentional terms, we make an advance over neo-Meinongianism to arrive at a more satisfactory neutralist account. This account is consistent with anti-realism because successful singular reference does not require that there be objects in the world which stand in some sort of relation to a speaker's words. Of course, sometimes objects answering to the designations we use may exist. But in a surprisingly large number of cases, as our earlier illustrations showed, there are no such objects. That does not stop us from talking about them or referring to them, for these activities are, at least in such cases, purely intentional activities.

The neutralist who is also an anti-realist will thus stand in the tradition of Brentano and Husserl in thinking of reference, not as a word-world relation, but as an intentional activity of agents which may or may not have a correlative real world object.[105] Brentano insisted upon the uniqueness of mental phenomena as object-directed or intentional. The object-directedness of mental reference does not imply that intentional objects exist in the external world; "All it means is that a mentally active subject is referring

104. Saul Kripke, "Speaker's Reference and Semantic Reference," in *Contemporary Perspectives in the Philosophy of Language*, P. French, Th. Uehling, and H. Wettstein (Minneapolis: University of Minnesota Press, 1979), pp. 6–27. As Searle notes, Kripke's distinction is not really about reference at all, since referring is something only speakers do, but is about the difference between the speaker's meaning and the linguistic meaning of the relevant expressions (John Searle, *Expression and Meaning: Studies in the Theory of Speech Acts* [Cambridge: Cambridge University Press, 1979], p. 155; idem, *Speech Acts,* Chapter 4). But the linguistic meaning of the words is beside the point so far as reference is concerned.

105. See Findlay, *Meinong's Theory of Objects*, pp. 40–41. As an illustration of a relational property Findlay gives *to-the-east-of-China*. If China is really a constituent of this property, then China must inhere, in all its solid immensity, in the Philippine Islands, which is absurd. Similarly, with respect to intentionality, if I think of *X*, my *thinking-of-X* is a mental relational property of which *X* is not a constituent. My state of mind is adequate to *X* because it has the relational property of *being-directed-to-X*. See also Margolis, "Reference as Relational," pp. 327–57; R. Scott Smith, *Naturalism and Our Knowledge of Reality: Testing Religious Truth-Claims* (Farnham, England: Ashgate, 2012), esp. Chapter 2, where he interprets Fred Dretske, Michael Tye, and William Lycan as endorsing a view of intentionality which construes it as a property, not a relation.

to them."[106] For Husserl, intentional objects are real world objects about which one is thinking; but not all intentional activity has intentional objects associated with it. Rejecting Meinongianism, Husserl held that when no object exists, the intentional activity exists without any object.[107] I can say things about non-existents like Pegasus, the accident that was prevented, holes, or numbers without committing myself to there being objects of which I am speaking. Thus, the non-existence of mathematical objects does not preclude our talking about or referring to them.

Summary

Free logic, neo-Meinongianism, and neutralism constitute formidable challenges to the criterion of ontological commitment that comes to expression in Premiss (I) of the Indispensability Argument. Construing reference as an intentional activity of agents and quantification as ontologically neutral accords with the data of ordinary language. By contrast taking singular reference to be ontologically committing and quantification to be existentially loaded not only hinders logical reasoning in important philosophical debates, but also falsifies much of what we believe or is outrageously **inflationary ontologically**, threatening to saddle us not only with innumerable abstract objects but fantasies of almost every sort.

106. Franz Brentano, "The Distinction between Mental and Physical Phenomena," trans. D. B. Terrell from *Psychologie vom empirischen Standpunckt* (1894), vol. 1, book 2, chapter 1, rep. in *Realism and the Background of Phenomenology*, ed. Roderick M. Chisholm (Atascadero, CA.: Ridgeview, 1960), pp. 50–51; idem, "Genuine and Fictitious Objects," trans. D. B. Terrell from *Psychologie vom empirischen Standpunckt*, Supplementary Essay IX (1911), rep. in *Realism and the Background of Phenomenology*, p. 71.

107. See Edmund Husserl, *Logical Investigations*, 2 vols. trans. J. M. Findlay [New York, NY: Humanities Press, 1970], 2: 595–6. For discussion, see Dallas Willard, *Logic and the Objectivity of Knowledge* (Athens: Ohio University Press, 1984), pp. 218–25; idem, "For Lack of Intentionality," in *Phenomenology 2005, Selected Essays from North America (Part 2)*, ed. Lester Embree and Thomas Nenon (Bucharest: Zeta Books, 2007), 5:593–612.

5. A Dispensability Argument

Geoffrey Hellman reflects that in the mid-twentieth century, it was common to regard Platonism in mathematics as the default position, since no non-Platonistic interpretation which preserved mathematical truth was available as a serious alternative. Now, however, the situation has changed. The Indispensability Argument "has been seriously weakened, if not yet quite completely vanquished."[108] The tables have sufficiently turned, such that it is not mere presumption to ask, "What, after all, is the *evidence* warranting belief in mathematical *abstracta*, numbers, sets, functions, etc.? Is there really any compelling or even good evidence?"[109] Hellman thinks that the answer is, "No," and our analysis of the Indispensability Argument bears this out this judgment.

Hellman, however, goes yet a step further to offer what Burgess and Rosen call a Dispensability Argument against Platonism.[110]

1. Synthetic claims of existence should not be given great credence without strong evidential support.
2. Existence claims for mathematical *abstracta*, especially the Axiom of Infinity, that there exists an infinite set or even plurality, are synthetic.
3. There is no good, direct evidence for such existence claims. For instance, we do not perceive sets, in any case not infinite sets, which classical analysis requires.
4. The best indirect evidence comes via scientific indispensability arguments. However, alternative programs such as Chihara's semantic modal-nominalism, and Hellman's modal-structuralism separately and cumulatively undercut such arguments.
5. The fact that current science takes mathematics for granted is not good evidence that premise (1) is suspended in the case of mathematics when viewed from a naturalistic-scientific standpoint.

---

108. Hellman, "On Nominalism," p. 693. In fact, Putnam himself came to embrace a sort of modal structuralism that preserves mathematical truth without mathematical objects (Hilary Putnam, "What Is Mathematical Truth?" in *Mathematics, Matter, and Method*, Philosophical Papers 1 [Cambridge: Cambridge University Press, 1979], pp. 60–78).
109. Hellman, "On Nominalism," p. 693.
110. Rosen and Burgess, "Nominalism Reconsidered," p. 519.

6. Therefore, the evidence currently available for mathematical *abstracta* is quite weak, and so we should not put much credence in a face-value, Platonist reading of mathematics.[111]

One might have added to (4) that the availability of neutral theories of reference and logic further undercut indispensability arguments.

In response to (4), Burgess and Rosen object that Hellman and Chihara's alternatives fail to undercut indispensability arguments because they are significantly less perspicuous, tractable, familiar, and fruitful for the further development of science. Such a complaint would, however, be significant only if the supposed "revolutionaries" were proposing their *ultima facie* interpretations as replacements for current theories. But they are not. Neither of them is advocating a scientific revolution in the name of anti-realism. They would agree that in light of standard theories' familiarity, utility, and so forth, we should go right on using the standard theories, while recognizing that we are not thereby committed to the existence of mathematical objects. Their alternatives are simply a way of meeting Quine's challenge to produce a mathematical theory adequate to the needs of modern science which does not involve quantification over or reference to abstract objects. The very fact that this challenge can be met in more than one way underlines the fact that these thinkers are not trying to put through a revolutionary mathematics but offering an undercutting defeater of the Indispensability Argument.

Burgess and Rosen interpret such anti-realists to claim that avoiding ontological commitment to mathematical objects outweighs the drawbacks of being less perspicuous, tractable, familiar, and so on. They question whether this consideration counts more than the drawbacks *by scientific standards*. Citing Maddy, they observe that historically, scientific theorists have been glad to proliferate mathematical objects. The revolutionaries are really promoting a *philosophical* standard of evaluation and therefore are not, after all, naturalized revolutionary nominalists.

Again, however, Burgess and Rosen misrepresent these *ultima facie* strategists. Since they are not advocating replacement of standard theories by anti-realist alternatives, they are not claiming that the admitted drawbacks of their formulations are outweighed by their

111. Hellman, "On Nominalism," p. 693.

fewer ontological commitments. Their claim is at most that the current theories, if taken prima facie and literally, involve Platonistic commitments, but that we are not committed to the reality of such objects by our pragmatic use of said theories, since they are capable of reinterpretation in such a way that the Platonic objects are dispensed with.

Whether these thinkers' objection to the literal truth of current theories is scientific or philosophical is difficult to answer because Burgess and Rosen provide no guidelines for what are scientific and what are philosophical standards of evaluation. Indeed, given Quine's so-called **naturalized epistemology**, it is difficult to make sense of such a bifurcation, since philosophy is supposed to be a part of science. For the naturalist, philosophical concerns about ontological economy will be broadly scientific. Maddy's point about natural science's cordiality toward new mathematical tools was precisely the reason she abandoned realism![112] If scientists thought that in adopting new mathematical apparatus, they were actually populating the world with entities as real as a new species of kangaroo or a hitherto unknown planetary system, then such proliferation of real entities would surely be a legitimate scientific concern and arouse serious scientific misgivings. In short, Burgess and Rosen have failed to turn back the force of the Dispensability Argument against mathematical Platonism.

Technically, however, Hellman's argument is not an argument for anti-Platonism but for agnosticism about Platonism. It justifies withholding belief in Platonism, not outright disbelief in Platonism. We might be tempted to strengthen (1) to read

> 1′. Synthetic claims of existence should be denied in the absence of evidential support.

The problem with (1′) is captured neatly by the aphorism, beloved of criminologists, that "Absence of evidence is not evidence of absence." For example, in theoretical physics, entities are frequently postulated for which there is (as yet) no evidence, but that absence of evidence in no way justifies one in thinking that such entities do not exist. Now clearly there are cases in which the absence of evidence *does*

112. See Penelope Maddy, "Believing the Axioms I," *Journal of Symbolic Logic* 53/2 (1988): 481–511.

constitute evidence of absence. If someone were to assert that there is an elephant on the quad, then the failure to observe an elephant there would be good reason to think that there is no elephant there. But if someone were to assert that there is a flea on the quad, then one's failure to observe it there would not constitute good evidence that there is no flea on the quad. The salient difference between these two cases is that in the one, but not the other, we should expect to see some evidence of the entity if in fact it existed. Thus, the absence of evidence is evidence of absence only in cases in which, were the postulated entity to exist, we should expect to have some evidence of its existence. The difficulty is that, given their causal isolation, abstract objects *cannot* leave evidence of their existence. So even if they did exist, we should not expect to have any evidence of them.

But perhaps their very causal isolation, due to their being causally effete, gives reason to deny their existence in the absence of any reason to believe in them. Van Inwagen explains,

> The platonist must think of objects, of what there is, as falling into two exclusive and exhaustive categories, the abstract and the concrete. If *x* falls into one of these categories and *y* into the other, then no two things could be more different than *x* and *y*. . . . the differences between God and this pen pale into insignificance when they are compared with the differences between this pen and the number 4; indeed, the number seems no more like the pen than like God. The difference between *any* abstract object and *any* concrete object would seem to be the maximum difference any two objects could display.[113]

Van Inwagen muses that it is very puzzling that objects should fall into two so radically different and exclusive categories as abstract and concrete. It would be much more appealing to suppose that one of the categories is empty. But concrete objects are indisputably real and well-understood, in contrast to abstract objects. So we should presume that abstract objects do not exist. Anti-Platonism of some sort is thus the default position. So, van Inwagen concludes, "one should not believe in abstract objects unless one feels rationally compelled by some weighty consideration or argument. . . . my conclusion is that a philosopher should wish not to be a platonist if it's rationally possible

113. Van Inwagen, "Theory of Properties," pp. 110–11.

for the informed philosopher not to be a platonist."[114] In fact, that conclusion is too soft: if, in the absence of countermanding argument, we ought to assume that the category of abstract objects is empty, then we should wish to be anti-Platonists and deny the existence of abstract objects. That opens the door to a Dispensability Argument.

Summary

Positive existence claims about such bizarre objects as abstract objects ought to be disbelieved in the absence of any overriding evidence or argument. The very dispensability of mathematical objects therefore provides grounds for thinking that such objects do not exist.

6. Concluding Remarks

The Indispensability Argument, which is the centerpiece of contemporary Platonism, is, in my studied opinion, a failure. Both its premises are plausibly false. But for the purposes of our debate, so bold an opinion need not be defended. In order to defeat the Indispensability Argument one need not rebut either of its premises but simply undercut the warrant for one of its premises. Given how controversial the debate over contemporary Platonism is, no view can plausibly commend itself philosophically to the exclusion of all others. As I have tried to show, there is a plethora of philosophically defensible anti-realist alternatives to Platonism on offer today. In order to undercut the Indispensability Argument, the anti-Platonist need not defend any one of these views but can recognize the viability of a plurality of viewpoints.

For my part, I find a combinatorial approach to abstract objects to be most plausible. To begin with, I think that neutralists are right that first-order logical quantification and use of singular terms are not ontologically committing. The first-order quantifiers should not be taken as devices of ontological commitment. A deflationary theory of reference affords us the option of construing reference, not as a word-world relation, but as an intentional activity of persons. As such, our use of singular terms need not commit us to the reality

114. Ibid., p. 107.

of what we refer to. Ordinary English seems to be a metaphysically lightweight language.

Of course, often we will find it convenient to speak of things like properties, propositions, numbers, and possible worlds. We can view such expressions as figurative or involving pretense. As I have argued, set theory seems to me a very plausible example of make-believe, whose affirmations carry no metaphysical weight.

Now on occasion we shall want to make our metaphysical commitments (or lack thereof!) clear. We shall want to say, for example, "There are no uncreated abstract objects." In this case, we are using the first-order quantifier to make a metaphysically heavyweight claim. That does not imply that we are speaking the imaginary language of Ontologese. We can make our metaphysical commitments clear in ordinary English by rhetorical devices or modifying phrases such as "There *really* are no uncreated abstract objects," or "There are no uncreated abstract objects in a metaphysically heavyweight sense of 'there are'."

Ontologese may be the language assumed by the fictionalist when he says that abstract object statements are untrue. But given a deflationary theory of reference, we need not go so far as the fictionalist in treating singular terms as devices of ontological commitment, even in Ontologese. Rather, with the positive free logician, we may take only the logical quantifiers of Ontologese to be ontologically committing. We can still affirm the truth of predications involving singular terms for things like numbers and properties, for example, "2 + 2 = 4" and "*Wisdom* is exemplified by Socrates" without implying "There is something which is identical to 4" or "There is something which is exemplified by Socrates," since in free logic the rules of Existential Generalization and Universal Instantiation apply only to things of which metaphysically heavyweight existence has been truly predicated, and we have not agreed or asserted that 2 + 2 exists or that *Wisdom* exists. Thus, one need not with the fictionalist deny the truth of such predications.

Such a combinatorial approach makes good sense of the many insights of the various anti-realisms. Almost all of them offer valuable insights, and we may glean from them the best of the lot.

First Round of Replies

Chapter 3

Reply to William Lane Craig's Opening Statement

Peter van Inwagen

Contents

1. Preliminary Remarks

My opening statement was in large part a reply to the arguments of Professor Craig's book *God Over All: Divine Aseity and the Challenge of Platonism* (2016) And Craig's opening statement was something very like a condensation of the arguments of *God Over All.* I think, therefore, that I have in effect already replied to his opening statement. I will therefore devote this "Reply" to an argument for a thesis about numbers—a kind of abstract object that was hardly mentioned in my opening statement, despite the title of this book. And that thesis is that if numbers and other mathematical objects do not exist, then the applicability of mathematics to the world about us (which cannot be disputed) is a mystery. The argument will not take a stand on a certain important question in the ontology of mathematics, namely: Are there *sui generis* numbers, things that are *intrinsically* numbers, objects that *must* be regarded as numbers—or are there only various classes of abstract objects, each of those very different classes comprising objects that can be made to do the work we want numbers to do? (Footnote 2 to my opening statement is relevant to this question.)

The argument of the present section will presuppose the meta-ontology defended in the opening statement. Of particular importance is my insistence that every statement in the following list entails each of

DOI: 10.4324/9781003008712-5

the others. (Which is not to say that they all *mean* exactly the same thing or that they are all interchangeable in every context.)

> Numbers exist
> Numbers exist in the strict and philosophical sense of 'exist'
> Numbers exist in the sense of 'exist' that pertains to ontological commitment
> Numbers exist in the metaphysically serious sense of 'exist'[1]
> $\exists x\ x$ is a number
> There is a number
> There are numbers
> Something is a number
> It is not the case that everything is something other than a number
> The number of numbers is not 0.

Before I turn to that task, however, I want to say something about one passage in Craig's opening statement, the summary of Section 4.1:

> The Platonist certainly has his work cut out for him if he is to provide adequate warrant for our affirming Premiss (II) of the Indispensability Argument. Wholly apart from proving the untenability of realist versions of anti-Platonism, he must defeat all of the defeaters offered by anti-realist alternatives to taking mathematical sentences to be literally true. Since the neo-Quinean himself sanctions the use of paraphrases in order to avoid unwanted ontological commitments, he faces the daunting task of showing that neither modal structuralism nor constructibilism successfully offers just such paraphrases. Moreover, he must show why we may not take mathematical discourse to be a kind of figurative language involving the use of existential metaphors, thereby avoiding commitment to mathematical objects. He must show why, despite common mathematical practice, axiomatic set theory may not be interpreted as proposing postulates which we are to imagine to be true without asserting them to be literally true. On top of it all, after all is said and done, the Platonist needs to explain why we cannot take the metaphysically heavy statements of mathematics,

1. Strictly speaking, my position is that *unless* this statement and the preceding two statements are nothing more than long-winded ways of saying 'Numbers exist', they are meaningless.

> Platonistically construed, to be simply false, however useful they may be in scientific practice and everyday life.

I don't know whether Craig meant this passage to imply that the Platonist must explicitly consider each of the things in Craig's long list of "defeaters offered by anti-realist alternatives to taking mathematical sentences to be literally true." If so, I can only say that I've spent a very long career refusing to let people tell me what I "must" do in philosophy and I don't propose to begin taking such direction now. (I'd rather take Craig's "Premiss II" apart and lay its pieces on the table and explain to anyone who wants to listen why I'd never use a premise put together from such mutually ill-adjusted parts.) As far as I can see, my presentation of my arguments for "lightweight platonism" in the opening statement pretty clearly display how I'd reply to any of the "defeaters" Craig mentions—generally by saying either, "That doesn't apply to my position" or "That's meaningless."

2. Reflections Upon the Word 'Three'

What explains the fact that applied mathematics is possible? What is the ground of the trust that we place in the answers to our questions about the world that calculation and mathematical reasoning provide? I will suggest that anyone who denies the existence of mathematical objects—sets, numbers, sequences, operators, functions, vectors, tensors, spaces—must regard the applicability of mathematics to the world as a mysterious, or at any rate an inexplicable, brute fact.

I will begin my defense of this suggestion by presenting some reflections on the role the word 'three' plays in a simple piece of mathematical reasoning—to wit:

> The committee has three members, and three is an odd number. Therefore, if every member of the committee votes, the vote won't be a tie—for if the number of people voting is odd, a tie vote is not possible.[2]

2. In all examples involving voting, it will be assumed that question the vote addresses is "binary": Smith or Jones; acquit or convict; aye or nay.

Note that the word 'three' "works" or "functions" differently in the sentences 'The committee has three members' and 'Three is an even number'. In the former sentence, 'three' is an adjective (modifying 'members'), and in the latter, 'three' is a noun (the subject of the sentence 'three is an odd number'). To see this, we need do no more than examine the results of substituting 'the number three' (obviously a nominal phrase and obviously *not* an adjectival phrase) for 'three' in those two sentences:

1. The committee has the number three members.

2. The number three is an odd number.

Sentence (2) is a perfectly meaningful sentence, and sentence (1) makes no sense at all. (If I were writing this in Latin—and you were reading it in Latin—it would be evident without argument that the occurrence of the Latin word for 'three' in the Latin version of 'the committee has three members' was an adjective, for it would agree in gender with the noun that followed it. 'Three members' is '*tria membra*' in Latin, but 'three persons' is '*tres personae*' and 'three human beings' is '*tres homines*'—because '*membrum*' is a neuter noun, and '*persona*' is feminine, and '*homo*' is masculine; and—in the nominative case—'*tres*' is used with masculine and feminine nouns and '*tria*' with neuter nouns.)

Although this is not customary usage, let us count English number-words—'six', 'forty-three', and so on—as "numerals." By an adjectival numeral, we understand a numeral that modifies a noun or noun-phrase (plural, of course—unless the numeral is '1' or 'one'). By a nominal numeral, we understand a numeral that can serve as the grammatical subject of a sentence. (In the sequel, we'll avoid the vaguely comic sound of the phrase 'nominal numeral' by abbreviating 'adjectival numeral' and 'nominal numeral' as 'adjectival' and 'nominal'.) In the remainder of this reply, adjectivals will be written in ordinary text, and nominals in bold-face, thus:

> If exactly three people vote, a tie vote is not possible because **three** is an odd number.

> The probability of a card drawn at random from a standard deck of 52 cards being a spade is 0.25, since 13 of the 52 cards are spades, and **13 ÷ 52 = 0.25**.

We may note that if we had only nominals in our language we could get along reasonably well in the world, and that this would not be the case if we had only adjectivals. If we had only nominals, we could not say "The committee has three members," but we could say, "The number of members of the committee is **three**," and that would do as well for practical purposes. But if we had only adjectivals we could not say "**Three** is an odd number"—and thus could not say, "Since there are three people voting, and **three** is an odd number, the vote cannot be a tie." One can, in fact, express what is expressed by our use of adjectivals without using any numerals at all. For example, instead of "The committee has three members," one could say.

> Someone *x* is a member of the committee, and someone *y* is a member of the committee, and someone *z* is a member of the committee, and no one else is a member of the committee and *x* is not *y*, and *x* is not *z*, and *y* is not *z*.

Summary

Section 1 is addressed to a methodological point that does not pertain to the main line of argument of this "Reply." In Section 2, it is shown that English "whole number" words (words like 'three' and 'forty-six') can be either adjectives or nouns. For example, 'three' is an adjective in 'There are three reasons to doubt what he has said,' and a noun in 'Three is a prime number'.

3. Applying Mathematics

It is the presence of nominals in our language that enables us not only to report the results of counting various things but also to refer to the "counting numbers" (integers, whole numbers) in abstraction from the counts of particular sets of objects, and, by means of such reference, to formulate general principles about the implications of facts about the numbers of things.

One such principle is:

> If the number of ballots cast is odd, a tie vote is not possible.

Call statements of the form 'There are *n* Xs' (where '*n*' represents the position of an adjectival and 'Xs' the position of a plural count-noun) simple numerical statements. The presence of nominals in our language allows us to state general principles that we can use to deduce important conclusions from simple numerical statements, conclusions that may or may not themselves be simple numerical statements. 'If x is the number of ballots cast and x is odd, a tie vote is not possible' is one such principle. Two more are:

> If the number of aspirin tablets in a bottle is x and the number of bottles in a crate is y, the number of tablets in a crate is $x \times y$.
>
> If the number of possible outcomes of a choice is x, and the number of favorable outcomes of that choice is y, and each possible outcome has the same probability, the probability of the choice's having a favorable outcome is $y \div x$.

It is upon such general principles that our ability to apply mathematics to the physical world depends. Our ability, that is, to expand by calculation the knowledge of the world that we get by counting and measuring things—to expand it by performing calculations with the numbers got by counting and measuring. I am going to present a fictional example of such a calculation, a very simple example, and ask why it would be rational to rely on the result of that calculation—and by extension, why it *is* rational to rely on the result of any correctly performed calculation. And here is my fiction:

> A certain farmer owns many sheep, so many that finding pasturage for them all is a constant source of worry for her. She has recently acquired a sizable rectangular field, and she wants to know how many of her sheep she will be able to keep in it. For reasons best known to agriculturists, she wishes the "stock density" of the field to be seven sheep per acre. She measured two adjacent sides of the field soon after she purchased it and found that they measured 170 meters and 380 meters. That evening, she wrote in her journal,
>
> > Today I performed the following calculation. The New Field is a 170-by-380-meter rectangle. Its area is therefore

> 64,600 square meters. An acre is 4047 square meters. The area of the New Field is therefore almost exactly 16 acres. I wish the stock density of the field to be 7 sheep per acre. I will therefore place 112 sheep in the New Field.

The "calculation" to which the farmer has referred is a fairly typical—albeit a very simple—exercise in applied mathematics. And most of us, I'm sure, would say that the procedures she employed (the procedure for determining the area of a rectangle given the lengths of two adjacent sides; the procedure for determining the population-density of a region, given its area and its population) were correct—that is, they could be trusted to give the right answer. But *why* do we trust these methods to give the right answer? Is it simply that they've never led us astray in the past? Is it simply that memory tells us that these methods—when applied correctly to reliable numerical data—have never led us to a pasture in which sheep were too densely packed or to a bridge that collapsed or the wrong date for an eclipse? (That was the position of the nineteenth-century English philosopher John Stuart Mill. (Mill, 1843/1967, Book II, Chs. IV, V, and VI.) Few have agreed with him.) If we say we trust these methods to give the right answer because and *only* because we know that they have faultless track records, however, our position implies that the trustworthiness of mathematics is a brute fact, a mystery. (If I ask, "Why is it reasonable to believe my car will start if I press the ignition button?" and the *only* answer anyone can give is, 'It always has', then no one knows why my car starts when I press its ignition button.) I think we can do better than that.

I would suggest that the trustworthiness of mathematics is best understood by thinking of each individual application of mathematics as a *piece of reasoning*. I would suggest that the best account of what our farmer was doing is this: she was reasoning her way from the premises.

3. The New Field is a 170-by-380-meter rectangle

and

4. An acre is 4047 square meters[3]

3. Close enough. The exact figure, if anyone is interested, is 4046.8564224.

to the conclusion,

5. If there are 112 sheep (i.e., 112 sheep and no other farm animals) in the New Field, the stock density of the New Field is 7 sheep per acre.

Let us call the argument whose premises are (3) and (4), and whose conclusion is (5) 'the Farmer's Argument'.

The Farmer's Argument is not logically valid. It is, however, possible to regard it as what is called an **enthymeme**—to suppose that it has suppressed (that is, not explicitly stated) premises—premises that were left unstated because to have stated them would have been pedantic. A simple example of an enthymeme, or at least of an argument that it's natural to regard as an enthymeme, is, 'This has eight legs, so it isn't an insect'—an argument that may be described as an "enthymematic abbreviation" of the logically valid argument.

This has eight legs

If this has eight legs, it isn't an insect [the "suppressed" premise of the enthymeme]

hence, It isn't an insect.

I propose to produce the logically valid argument of which the Farmer's Argument is an enthymematic abbreviation—or, more cautiously, I propose to produce *a* logically valid argument such that it's at least plausible to regard the Farmer's Argument as an enthymematic abbreviation of that argument.

I propose to do this because, in my view,

(a) The existence of numbers is not a logical consequence of (3) and (4).

(b) Although the Farmer's Argument is not logically valid, it is certainly an unexceptionable piece of reasoning; this is because it is an enthymematic abbreviation of a valid argument whose suppressed premises are—or at least would be regarded by many as—self-evidently true.

(c) Some among those suppressed premises imply the existence of numbers.

Before I state the argument of which the Farmer's Argument is an enthymematic abbreviation, however, I shall have to introduce a new piece of terminology, and this will have to be done with some care.

When we use adjectivals and nominals we sometimes use them to say how many things of some specified sort there are (sheep in a certain field, electrons in the outer shell of the boron atom, prime numbers between 100 and 700, etc.). For example:

> There are 100 sheep in the New Field
> The number of sheep in the New Field is 100.

> There are 100 prime numbers between 100 and 700
> The number of prime numbers between 100 and 700 is 100.

But we use numerals not only to enumerate but to report the results of measurements. (And a larger class of numerals—for the result of a measurement is generally not a "whole" number. But this fact, important as it is for the philosophy of applied mathematics, will not figure in my argument in any essential way.) Suppose that I (having measured one side of the New Field and having weighed our sheep Boston Blackie) say these things:

> The New Field is 380 meters in length

and

> Boston Blackie weighs 140 kilograms.[4]

How shall I phrase the corresponding nominal statements? What sentence stands to 'The New Field is 380 meters in length' as 'The number of sheep in the New Field is 100' stands to 'There are 100 sheep in the New Field'? (If I cannot answer this question, I shall not be able to state the suppressed premises of the argument of which the Farmer's Argument is an enthymematic abbreviation.)

We can, I suppose, say—just barely—"The number of kilograms Boston Blackie weighs is 140," but there is no real point in going

4. Recall this convention: In everyday, non-scientific contexts, in discourse about objects on or close to the surface of the earth, '*x* weighs *N* kilograms' abbreviates 'the weight of *x* is the weight of an object on the surface of the earth whose mass is *N* kilograms'.

down that road, for the plain fact is that the statement 'Boston Blackie weighs 140 kilograms' reports not a count but a measurement. If someone says, "Boston University has 140 student organizations," the reply (perhaps by some state official), "I'm going to need a list of them" would make sense. And one would respond to that demand by providing a list of student organizations that had 140 entries. But if someone says, "Boston Blackie weighs 140 kilograms," the response, "I'm going to need a list of them"[5] makes exactly as much sense as the idea of a list of kilograms that has 140 entries—which is to say, no sense at all. Or, again, if someone says, "There were 140 guests at each of their two largest parties last year," it makes sense to say, "I wonder if it was the same 140 guests" in reply; but if someone says, "The two largest sheep they had last year each weighed 140 kilograms," it makes no sense at all to say, "I wonder if it was the same 140 kilograms" in reply. Kilograms are not objects—or perhaps it would be better to say that no object can be correctly called "a kilogram." (Well, the so-called prototype kilogram or ur-kilogram is an object, but it doesn't figure in the weight or the mass of anything: one can't say, 'Boston Blackie weighs 140 kilograms, and the prototype kilogram is one of them'.) And "objects" in the present sense—the very abstract sense employed by mathematicians and philosophers—are precisely those things that can be counted. 'Boston Blackie weighs 140 kilograms' is a statement of the result of measuring a sheep's weight, not a statement of the size of a population of kilograms. (And, in any case, a sheep that weighs 140 kilograms today might weigh 140.4462 kilograms tomorrow. No university has 140.4462 student organizations; there have never been 140.4462 guests at a party.)

Our question is, how shall we phrase the nominal statements that "parallel" the adjectival statements that report the results of measurements (as opposed to counts)? I propose to phrase them by using a form of words invented by the twentieth-century American philosopher W. V. O. Quine,[6] a form illustrated by these two 'if and only if' statements:

> **380** is the length-in-meters of the New Field if and only if the New Field is 380 meters in length

---

5. "What?—a list of our sheep?" "No, no—a list of the kilograms your sheep Boston Blackie weighs."
6. Compare Quine, 1960, 245.

> **140** is the weight-in-kilograms of Boston Blackie if and only if Boston Blackie weighs 140 kilograms.

(Cf., '**112** is the number of sheep in the New Field if and only if there are 112 sheep in the New Field'.)

In the loose sense in which the sentence 'Milan is south of Denmark' states that a certain city stands in a certain relation to a certain nation, '**380** is the length-in-meters of the New Field' states that a certain number stands in a certain relation to a certain field. There are many relational phrases that perform functions analogous to those of 'is the length-in-meters of' and 'is the weight-in-kilograms of': 'is the height-in-cubits of', 'is the age-in-seconds of', 'is the volume-in-cubic-light-years of', 'is the density-in-grams-per-cubic-centimeter of' . . . Such phrases are two-place relational predicates, predicates of the same sort as 'is the color of'—as it occurs in the sentence 'Black is the color of my true love's hair'. (I am assuming that here 'black' is a noun, the subject of the predicate 'is the color of my true love's hair'.) In that sentence, the predicate 'is the color of' relates a certain abstract object, a color, the color black, to a certain concrete object, my true love's hair. In the sentence '**140** is the weight-in-kilograms of Boston Blackie', the predicate 'is the weight-in-kilograms of' likewise relates a certain abstract object to a certain concrete object: it relates a number, the number **140**, to a certain living organism, Boston Blackie.

We may now state explicitly a logically valid argument of which the Farmer's Argument can be supposed to be an enthymematic abbreviation. (In this argument, the premises (6), (7), and (9) are roughly equivalent to (4) and (5) in the Farmer's Argument.)

6. The New Field is a rectangular region

    Premise

7. The length of one side of the New Field is 170 meters, and the length of an adjacent side of the New Field is 380 meters

    Premise

8. For any *x*, if *x* is a rectangular region and the length of one side of *x* is 170 meters, and the length of an adjacent side of *x* is 380 meters, then **170** is the length-in-meters of one side of *x*, and **380** is the length-in-meters of an adjacent side of *x*

    Premise

9. For any x, y, z, and w, if x is a rectangular region, and if y is the length-in-meters of one side of x, and z is the length-in-meters of an adjacent side of x, and if $w = (y \times z) \div 4047$, then w is the area-in-acres of x

Premise

10. (**170** × **380**) ÷ **4047** = **16**[7]

Premise

11. **16** is the area-in-acres of the New Field

From 6 to 10

12. For any x, any y, and any z, if x is a rectangular region, y is the area-in-acres of x, and z is the number of sheep in x, then $z \div y$ is the stock-density-in-sheep-per-acre of x

Premise

13. If **16** is the area-in-acres of the New Field, and **112** is the number of sheep in the New Field, then **112** ÷ **16** is the stock-density-in-sheep-per-acre of the New Field

From 6, 12

14. **112** ÷ **16** = **7**

Premise

15. If **112** is the number of sheep in the New Field, then **7** is the stock-density-in-sheep-per-acre of the New Field

From 11, 13, 14

16. **112** is the number of sheep in the New Field if and only if there are 112 sheep in the New Field

Premise

17. **7** is the stock-density-in-sheep-per-acre of the New Field if and only if the stock density of the New Field is 7 sheep per acre

Premise

5. If there are 112 sheep in the New Field, the stock density of the New Field is 7 sheep per acre.

From 15, 16, 17

7. The actual quotient is more like 15.962441314553991... or if the strictly correct square-meters-to-acres divisor (4046.8564224; see footnote 3) had been used, more like 15.963007642778881... I will ignore these refinements.

If this argument—"the Farmer's Argument (Expanded)," let us call it—is sound (that is, if it is logically valid and has true premises), there is an answer to the question, 'Why was the Farmer's reasoning trustworthy?'. And that answer is

> The Farmer's Argument is an enthymematic abbreviation of the Farmer's Argument (Expanded), which is a sound argument. That the figure the farmer's calculation produced was correct has, and needs, no more explanation than this: The conclusion of a sound argument must be true.

One might of course go on to ask, "But what about the premises of the argument (other than (6) and (7))? What about propositions (8), (9), (10), (12), (14), (16), and (17)? Why should one accept those?" And that's a fair question. Its fairness shows that our answer is at best incomplete, at best, only the beginning of a complete answer. (I'll continue to call it an "answer," however.) But note: anyone who thinks that any of those premises is false cannot accept an answer that includes the statement that the Farmer's Argument is an enthymematic abbreviation of the Farmer's Argument (Expanded).

4. Nominalism and Applied Mathematics

Is this answer available to the nominalist? One possible response to this question is, "Of course not, for its premises are inconsistent with, or have consequences that are inconsistent with, nominalism. Nominalism entails the falsity of several of the premises; (11) and (14) among them." This should be obvious: (11) entails

$\exists x$ x is the area-in-acres of the New Field

and (14) entails

$\exists x\ x \div \mathbf{16} = 7$,

and the **open** sentences 'x is the area-in-acres of the New Field' and '$x \div \mathbf{16} = 7$' can be **satisfied** only by entities whose existence would be inconsistent with nominalism.

But that response would be a bit hasty. It is true that '**16**' and '**112**' are nominals, but it does not follow that the positions they occupy are subject to existential generalization.

Some Terms from Logic

Existential generalization is a rule of inference two examples of the application of which are

Socrates was wiser than Protagoras, *and therefore*
Something^x was such that (it^x was wiser than Protagoras) [*or* '$\exists x$ x was wiser than Protagoras']

The square root of 2 is an irrational number, *and therefore*
Something^x is such that (its^x square root is an irrational number).

In the Opening Statement, we encountered the idea of a phrase that does (or does not) "occupy a position subject to existential generalization." Here I will say a bit more about this idea. We may sometimes be convinced that a noun or noun-phrase within a sentence does not contribute to the meaning of that sentence by denoting an object. For example:

Santa Claus does not exist
There are 2.53 residents in the average American household.

After all, if 'Santa Claus' denoted something, it would denote Santa Claus—and Santa Claus *would* exist. In this case, we say that 'Santa Claus' does not occupy a position subject to existential generalization—for in such cases, 'Something^x is such that (it^x does not exist)' does not follow from 'Santa Claus does not exist'. And, although it is true that there are 2.53 residents in the average American household, it does not follow that $\exists x$ there are 2.53 residents in x. That is to say, in that sentence 'the average American household' does not occupy a position subject to existential generalization.

A sentence that contains occurrences of variables that are not paired with any occurrence of a quantifier (a so-called **free** variable) is said to be an **open** sentence. We say that an object **satisfies** an open sentence if it satisfies the condition the sentence expresses. The set of things that satisfy an open sentences is its **extension**. So, for example, 'x was wiser than Protagoras' is an open sentence, the sole occurrence of 'x' in that sentence is free, the sentence is satisfied by those and only those things that are wiser than Protagoras, and its extension is the set of things that are wiser than Protagoras.

And nor does it follow that (for example) proposition (11) ('**16** is the area-in-acres of the New Field') is a subject-predicate proposition: it does not follow that '**16**' denotes a certain number, a certain abstract object, and that the predicate 'is the area-in-acres of the New Field' ascribes a certain property to that number. True, that is how a Platonist will understand (11), but it does not follow from simply the fact that '**16**' is a nominal that the Platonist is right. After all, 'the shadow of the flagpole' is a noun phrase, and, for all I know, the proposition expressed by

18. The shadow of the flagpole was twice as long at 4 p.m. as it was at noon

is true, and if it *is* true, it does not follow from its truth that

19. $\exists x$ x was twice as long at 4 p.m. as it was at noon.

For, in my view, (18) could be true even if 'the shadow of the flagpole' denotes nothing and nothing belongs to the extension of 'x was twice as long at 4 p.m. as it was at noon'. And if that is so, '**16** is the area-in-acres of the New Field' might be true even though '**16**' denotes nothing and nothing belongs to the extension of 'x is the area-in-acres of the New Field'.

But now let us consider premise (12),

> For any x, any y, and any z, if x is a rectangular region, y is the area-in-acres of x, and z is the number of sheep in x, then $z \div y$ is the stock-density-in-sheep-per-acre of x.

(The points I shall make about (12) could also be made about premise (9).) Premise (12) is logically consistent with nominalism—in fact, nominalism *entails* (12) (just as 'There are no unicorns' entails 'All unicorns have one horn'). Unfortunately, nominalism entails an infinite number of "useless rivals" to (12), such as

> For any x, any y, and any z, if x is a rectangular region, y is the area-in-acres of x, and z is the number of sheep in x, then $\mathbf{13}z \div \mathbf{6.57}y$ is the stock-density-in-sheep-per-acre of x

and

> For any x, any y, and any z, if x is a rectangular region, y is the area-in-acres of x, and z is the number of sheep in x, then $z \div y$ is the stock-density-in-sheep-per-square-parsec of x.

('There are no unicorns' entails 'All unicorns have one horn', but it also entails 'All unicorns have no horns', 'All unicorns have seventeen horns', and 'All unicorns have bank accounts.') Nominalists who use (12) owe us an explanation of why they use (12) rather than one of the "useless rivals" (all of them true, and true for the same reason as (12)). But I will not press this point, because in my view, (12) confronts the nominalist with a more fundamental difficulty than this. (Premise (9) raises essentially the same difficulty.) Note that line (13) is said to follow from (6) and (12). If it does indeed follow from (6) and (12), it follows by three applications of Universal Instantiation.

Another Term from Logic

Universal Instantiation is a rule of inference. Here is an example of an application of this rule:

> All Greeks are mortal, *and therefore*
> If Socrates is a Greek, Socrates is mortal

Or, in Revised English,

> Everything^{x} is such that (if it^{x} is a Greek, then it^{x} is mortal), *and therefore*
> If Socrates is a Greek, Socrates is mortal

Or, in our abbreviated Revised English

> $\forall x$ (if x is a Greek, x is mortal), *and therefore*
> If Socrates is a Greek, Socrates is mortal.

But if

13. If **16** is the area-in-acres of the New Field, and **112** is the number of sheep in the New Field, then **112 ÷ 16** is the stock-density-in-sheep-per-acre of the New Field

follows from premise (12) by three applications of Universal Instantiation (and a small assist from premise (6)), then

21. $\exists x$ x is the area-in-acres of the New Field

follows from

11. **16** is the area-in-acres of the New Field

by Existential Generalization. There is a perfectly general principle at work here. Suppose, for example, that someone believes that

22. The shadow of the flagpole is a physical thing

follows logically from

23. $\forall x$ (x is visible, then x is a physical thing)

and

The shadow of the flagpole is visible.

Then that person should also believe that '$\exists x$ x was twice as long at 4 p.m. as it was at noon' follows from 'The shadow of the flagpole was twice as long at 4 p.m. as it was at noon'. As we have said, someone may certainly maintain that 'The shadow of the flagpole was twice as long at 4 p.m. as it was at noon' is true and that '$\exists x$ x was twice as long at 4 p.m. as it was at noon' is false, but anyone who does maintain that cannot also maintain that, e.g., 'the shadow of the flagpole is a physical thing' follows logically from '$\forall x$ (x is visible, then x is a physical thing)' and 'The shadow of the flagpole is visible'.

It is not hard to see why this "general principle" holds true. Consider, for example, proposition (23).

This is, essentially, the proposition that

The things that there are are (one and all) physical things if they are visible.

And (22) follows from that proposition and 'The shadow of the flagpole is visible' only if 'the shadow of the flagpole' is the name of one of the things that there are. (Of course, any word or phrase that names anything at all must name of one of the things that there are. Therefore, if 'the shadow of the flagpole' is *not* a name of one of the things that there are, it is not a name of anything at all. And that in fact is the position of those who hold that one cannot deduce '$\exists x$ x was twice as long at 4 p.m. as it was at noon' from 'The shadow of the flagpole was twice as long at 4 p.m. as it was at noon': owing to the fact that the nominal 'the shadow of the flagpole' is not a name of anything at all, it does not occupy a position subject to existential generalization.) And if 'the shadow of the flagpole' *is* a name of one of the things that there are, then—obviously—'∃*x* *x* was twice as long at 4 p.m. as it was at noon' follows from 'The shadow of the flagpole was twice as long at 4 p.m. as it was at noon'.

There are many strategies by which nominalists might undertake to resist my contention that nominalism entails that the Farmer's Argument (Expanded) has false premises. But suppose they concede that that entailment holds. Suppose, in fact, that they concede further that the fact that entailment holds leaves them without an account of the trustworthiness of our imaginary farmer's calculation (her calculation of the number of sheep that she must place in the New Field if she wishes to achieve a stock-density of 7 sheep per acre). Suppose they tell me that, although the Farmer's "calculation," namely

> The New Field is a 170 by 380 meter rectangle. Its area is therefore 64,600 square meters. An acre is 4047 square meters. The area of the New Field is therefore almost exactly 16 acres. I wish the stock-density of the field to be 7 sheep per acre. I should therefore place 112 sheep in the New Field.

is obviously an unexceptionable application of mathematics to the physical world, they must admit that they can offer no argument or other general consideration in support of this judgment—or none but this: the Farmer is applying methods that seem always to yield the right answers. What would be wrong with their simply conceding those things?

Summary

Sections 3 and 4 address the question, why do we rely on the results of numerical calculations? If we look at a calculation as a piece of reasoning, as an argument whose premises comprise certain numerical statements, and whose conclusion is a numerical statement, then our question is answered if we know that such arguments are logically valid and have true premises—for if we know the premises of an argument we know to be logically valid to be true, we shall, of course, trust its conclusion. Analysis of the logical structure of a typical calculation shows that its premises can be true only if numbers exist.

5. Nominalism and Millianism

Nothing, really. Except that they are thereby committed to accepting a "Millian" epistemology of applied mathematics. To see why this is so, let us consider an application of mathematics to the physical world that is even simpler than the "sheep in the field" calculation, namely to the problem of calculating the number of aspirin tablets in a crate if a crate contains 8000 bottles and each bottle contains 500 aspirin tablets.

"Well, that's too simple even to be called a problem. There are the same number of aspirin tablets in each bottle, so just multiply the number of bottles by that number."

I agree, of course. But let's ask *why* that procedure yields the right answer. In the present case, the right answer is 'The number of aspirin tablets in a crate is 4,000,000'. And that answer (*I* would say) is obtained by first taking the general rule (which you—you the Interlocutor—in effect stated, but here it is put a bit more formally than you did):

24. $\forall x \forall y$ (if the number of bottles in a crate = x, and the number of aspirin tablets in a bottle = y, then the number of aspirin tablets in a crate = $x \times y$),

and then applying the logical rule of Universal Instantiation twice to (24), to obtain

25. If the number of bottles in a crate = 8000 and the number of aspirin tablets in a bottle = 500, then the number of aspirin tablets in a crate = 8000 × 500.

And, finally, deducing

26. The number of aspirin tablets in a crate = 4,000,000

from (25) and two premises given in the statement of the problem,

The number of bottles in a crate = 8000

The number of aspirin tablets in a bottle = 500

and the premise (supplied by arithmetic)

27. 8000 × 500 = 4,000,000.

As we have seen, however, that can't be how "that answer is obtained" by the *nominalist*. Not, at any rate, if the nominalist is a consistent nominalist. For if (25) is to follow from (24) by Universal Instantiation, then the nominals '8000' and '500' must be so understood that statements inconsistent with nominalism are deducible from (26) and (27) by Existential Generalization. For example, if (25) follows from (24) by Universal Instantiation,

$\exists x$ the number of aspirin tablets in a crate = x

follows from (26) by Existential Generalization. And the number of aspirin tablets that satisfy any given condition is a universal. (For all I know, the number of aspirin tablets that are in my house as I write these words is also the number of Albanians named Belushi and the number of asteroids inside the heliosphere with masses of more than one million metric tons.)

How then might a consistent nominalist approach the "aspirin" problem? Perhaps in some such way as this:

> Instead of general principles that are universal quantifications over numbers—that is, principles like (24)—use sentential *schemata* in which the schematic letters represent adjectivals. For example, in the present case, the schema might be this:
>
> *S* If there are *N* bottles in a crate and *n* aspirin tablets in a bottle, there are $\mathcal{N}$ aspirin tablets in a crate.

And the rule for using *S* would be:

> R_S Any sentence that comes from S by the substitution of adjectivals for '*N*' and '*n*' and the substitution for '$\mathcal{N}$' of the result of applying the multiplicative algorithm for adjectivals to the adjectivals that were substituted for '*N*' and '*n*' is true.

Let us call this suggestion the "Suggested Way." By the "multiplicative algorithm for adjectivals," advocates of the Suggested Way understand the mechanical techniques of calculation that we all learned in the golden years of our youth—the techniques productive of displays like

$$\begin{array}{r} 380 \\ \times\, 170 \\ \hline 26600 \\ 380 \\ \hline 64600. \end{array}$$

Applied to the particular "aspirin problem" we have been considering, the Suggested Way yields

> 'If there are 8000 bottles in a crate and 500 aspirin tablets in a bottle, there are 4,000,000 aspirin tablets in a crate' is true,

which obviously entails

> If there are 8000 bottles in a crate and 500 aspirin tablets in a bottle, there are 4,000,000 aspirin tablets in a crate

—a proposition that differs from

> If the number of bottles in a crate = 8000 and the number of aspirin tablets in a bottle = 500, the number of aspirin tablets in a crate = 4,000,000

only in ways that are of no practical interest, however interesting they might be to a few philosophers.

Now it may be objected that the use of the Suggested Way involves one no less in ontological commitment to abstract objects that does the use of quantificational principle (24). Note that while *S* is indeed a schema and not a sentence, R_S, the "rule for using *S*" is not a schema but a sentence. Its content may be expressed a little more explicitly in this way:

28. $\forall x \forall y \forall z$ (if *x* is an adjectival and *y* is an adjectival and *z* is a sentence that comes from the schema 'If there are *N* bottles in a crate and *n* aspirin tablets in a bottle, there are $\mathcal{N}$ aspirin tablets in a crate' by the substitution of *x* for '*N*' and *y* for '*n*' and the result of applying the multiplicative algorithm for adjectivals to *x* and *y* for '$\mathcal{N}$', then *z* is true).

And if we wish to use (28) to determine the number of aspirin tablets in a crate, given that there are 8000 bottles in a crate and 500 tablets in a bottle, we must believe that

29. If '8000' is an adjectival and '500' is an adjectival and 'If there are 8000 bottles in a crate and 500 aspirin tablets in a bottle, there are 4,000,000 aspirin tablets in a crate' is a sentence that comes from the schema 'If there are *N* bottles in a crate and *n* aspirin tablets in a bottle, there are $\mathcal{N}$ aspirin tablets in a crate' by the substitution of '8000' for '*N*' and '500' for '*n*' and the result of applying the multiplicative algorithm for adjectivals to '8000' and '500' for '$\mathcal{N}$', then 'If there are 8000 bottles in a crate and 500 aspirin tablets in a bottle, there are 4,000,000' aspirin tablets in a crate' is true

follows from (28) by Universal Instantiation. And we have, in effect, seen that if (29) follows from (28) by Universal Instantiation, then

$\exists x$ *x* is an adjectival

follows from

'8000' is an adjectival

by Existential Generalization. And adjectivals are abstract objects, universals. Adjectivals are no more acceptable to nominalists than numbers or attributes or any other kind of abstract object. Adjectivals are adjectives, and adjectives are words, and words are universals. If you doubt this, consider these sentences, which it took me only a few seconds to find on line:

There are about 8000 species of ant
About 8000 general practitioners have X-ray equipment
The U.S. imports about 8000 tons of hemp every year.

The same adjectival, '8000', occurs in all three statements. Some medieval nominalists would have said that that the adjectival '8000' was a *flatus vocis*, a mere pulse of air through someone's vocal cords, a mere puff of sound. But they were wrong: the adjectival '8000' is not a mere puff of sound—although it is what is common to millions and millions of mere puffs of sound.

If this argument is right, nominalists cannot consistently avail themselves of the Suggested Way. There are, of course, few if any knock-down arguments in philosophy, and there are various ways in which nominalists might dispute this one. Some of these ways are on the level of meta-ontology. One might, for example, affirm, with Craig and his allies, that nominalists can consistently say both 'There are no abstract objects' and 'There are abstract objects'—using the words 'there are' in an ontologically serious way in the former but not the latter assertion.

Is it possible so to modify the Suggested Way that the modified version does not involve those who make use of it in "ontological commitment" to abstract objects? For the remainder of this reply, by 'a nominalist' I will mean 'a philosopher' who denies the existence of abstract objects and who is an adherent of a neo-Quinean meta-ontology—that is, a meta-ontology according to which the sentences

Abstract objects exist
Abstract objects exist in the strict and philosophical sense of 'exist'
Abstract objects exist in the sense of 'exist' that pertains to ontological commitment
Abstract objects exist in the metaphysically serious sense of 'exist'[8]
$\exists x$ x is an abstract object
There is an abstract object
There are abstract objects
Something is an abstract object
It is not the case that everything is something other than an abstract object
The number of abstract objects is not 0

are equivalent statements—different ways of saying *essentially* the same thing (as 'All Greeks are mortal' and 'There are no immortal Greeks' are two ways of saying essentially the same thing.)

Anyone who accepts a neo-Quinean meta-ontology and who denies the existence of abstract objects (numbers among them—and adjectivals among them as well) must affirm the following thesis:

Consider sentences of the form:

$\exists x\ (\ldots x \ldots x \ldots)$

8. Strictly speaking, my own position is that *unless* this statement and the preceding two statements are nothing more than long-winded ways of saying 'Abstract objects exist', they are meaningless.

> —where '...*x*...*x*...' represents a sentence in which '*x*' and no other variable is free. No sentence of that form is true if the open sentence represented by '...*x*...*x*...' could be satisfied only by abstract objects. (For example, '∃*x* the number of aspirin tablets in a crate = *x*' and '∃*x x* is an adjectival' are not true, because 'the number of aspirin tablets in a crate = *x*' and '*x* is an adjectival' can be satisfied only by abstract objects.)

Can a neo-Quinean nominalist adopt the Suggested Way? Not—obviously—if adopting the Suggested Way involves accepting the existence of adjectivals. But consider my colleague Norma—a neo-Quinean *and* a would-be nominalist. Norma thinks she sees a trick that will enable her to adopt the Suggested Way without being thereby forced to accept the existence of adjectivals (or abstract objects of any other kind). Her reasoning is as follows:

> Others who adopt the Suggested Way may do so partly because they *believe* certain things—that (27) is true, for example, or that '8000' is an adjectival. Well, *I'm* going to adopt the Suggested Way without believing any of those things. I can do that because I know how those who *do* believe these things *act*; and, in particular, I know what marks they make on paper when, as they put it, they are "multiplying a number by another number" and "dividing one number by another." But one can make those marks without accepting any proposition whose truth entails the existence numbers or of adjectivals or of any other abstract object. And that's what I'll do. I'll act *as if* I believed certain things without actually believing them. I will adopt the position that German philosophers around the turn of the twentieth century called "the philosophy of 'as if'" and present-day anglophone philosophers call fictionalism.

Strictly speaking, the Suggested Way applied only to problems of the form, 'How many aspirin tablets are there in a crate, if there are so-and-so many tablets in a bottle, and such-and-such many bottles in a crate'. But the idea behind it is obviously easily generalized. So, for example, if our farmer adopted the analogue of the Suggested Way for her problem (and applied it *à la* Norma), a record of the steps she followed in her calculation would look something like this:

The New Field is a 170-by-380-meter rectangle. Its area is therefore [at this point she makes the following marks on a sheet of paper]

$$\begin{array}{r} 380 \\ \times\ 170 \\ \hline 26600 \\ 380 \\ \hline 64600 \end{array}$$

64,600 square meters. An acre is 4047 square meters. The area of the New Field is therefore [at this point she makes the following marks on a sheet of paper]

$$\begin{array}{rl} & 15.96 \\ 4047 & \overline{\big)\,64600.00} \\ & \underline{4047} \\ & 24130 \\ & \underline{20235} \\ & 38950 \\ & \underline{36423} \\ & 25270 \\ & \underline{24282} \\ & 9880\ldots \end{array}$$

almost exactly 16 acres. I wish the stock density of the field to be 7 sheep per acre. I will therefore [at this point she makes the following marks on a sheet of paper]

$$\frac{x}{16} = \frac{7}{1}$$

$$x = 16 \times 7 = 112$$

place 112 sheep in the New Field.

And when she is inscribing those three groupings of "marks on a sheet of paper," she is following no rule—or at any rate she accepts no proposition that would justify or endorse those three displays of marks (that is, no proposition like (27)). If she is following rules, it is in the rather abstract sense in which a computer "follows rules." She behaves this way—makes these marks and allows the marks she has made to influence her beliefs, allows them in fact to cause her to believe such things as '112 sheep is the number of sheep to place in the field to achieve a stock density of 7 sheep per acre'—because she has been trained to behave in that way. If you like, she and her pencil and paper are parts of a calculating machine—a machine made of flesh and blood and cellulose and graphite. The programming of the machine is a set of procedures for making sequences of pencil marks on paper that one of its components, its human "core processor," has somehow internalized.[9]

And if our farmer follows Norma's example and so behaves (in all situations that require the processing of quantitative information), she will believe, or will at any rate *need* to believe, nothing inconsistent with nominalism.

She may, however, find herself with questions. Suppose it occurs to her one day to wonder whether these techniques she uses, these various procedures for making sequences of pencil marks on paper, are reliable—whether the beliefs they lead her to—that she should put 112 sheep into the New Field to achieve a stock density of 7 sheep per acre, and the like—are in every case true. Platonists who ask that question can be shown a proof that (for example) the marks-on-paper techniques for manipulating adjectivals displayed above never yield a falsehood when applied to truths. That is to say, platonists can be shown what they will recognize as a proof of that proposition, but nominalists who are shown this supposed proof will protest that it is not a proof because it rests on false premises. Nominalists will agree, one and all, with this statement:

> The argument you have offered us as a proof has premises that can be true only if certain open sentences are satisfied.[10] And those sentences could be satisfied only by numbers. And there *are* no numbers, for everything there is is a concrete material particular and hence not a number.

It would seem, therefore, that the only reasons nominalists could have for trusting the rules for making marks on paper that their elementary-school teachers have trained them to follow are empirical: carefully maintained records of vast numbers of applications of these procedures demonstrate that following them has never led anyone astray. Suppose, for example, that you put 23 aspirin tablets into each of 116 bottles, 91 tablets into each of 48 bottles, and 253 tablets into each of 11 bottles, and count the total number of tablets involved in each of

---

9. Compare Asimov, 1958.v.
10. Well, they will agree, one and all if they don't consider the question whether the statement entails the existence of open sentences.

these distributions. You will find that in each case, your count matches the numeral at the foot of one of these three columns of **marks**:

23	91	253
×116	×48	×11
138	728	253
23	364	253
23	4368	2783
2668		

And, of course there will be a *vast* amount of similar evidence for the reliability of these procedures, so much that no one could doubt their reliability. Undeniable as the fact of their reliability may be, however, no nominalist will be able to explain it: it will be for them the brutest of facts. (Of course, the nominalist will insist that neither do the platonists have an explanation of the reliability of the algorithms of elementary-school arithmetic: "Oh, to be sure, they have an explanation. The only problem with it is, it's wrong.")

It is for this reason that I have said that nominalists are committed to accepting a "Millian" epistemology of applied mathematics. And, for my part, I make bold to say that Mill's epistemology of mathematics is obviously wrong. If Mill is right, mathematics is a sort of black box that supplies answers to property framed questions about the implications of data gathered by counting and measuring. If Mill is right, it's rather as if each of us had a literal black box, something like a smart phone, which the box's owner addresses as 'Mathematica' (cf. Siri and Alexa). The following dialogue is an example of a typical occasion of the application of a box (in this case by an aerospace engineer):

"Hey, Mathematica!"

"How can I help you?"

"If there's a toroidal space station 200 meters in diameter, what should the angular velocity of its rotation be if astronauts standing on inner surface of its outer wall are to be subject to a centripetal acceleration exactly equal to the acceleration due to gravity at the surface of the earth. State the answer in revolutions per minute, please."

"Is '9.8 m/sec^2 an acceptable value for the acceleration due to gravity at the surface of the earth?"

"Yes."

"The answer to your question is 2.9894 rpm."

As I say, everyone in our imaginary world has such a box, and everyone trusts the answers the boxes give to the questions they're asked. (Remember, the only questions they'll answer are the ones whose answers could in principle be found by human beings wielding pencils and paper and slide rules and log tables and such.) And the reason they trust those answers is that they're always right. But no one has the faintest idea how the boxes work. If Mill's epistemology of mathematics is right then each applied mathematician (or each applied mathematician taken together with certain "equipment," such as pencil and pieces of paper) is like one of those black boxes—a calculator made of flesh and blood and bone and cellulose and graphite.

A nominalist must be content with this conclusion. Since I could never be, I am not a nominalist.

Summary

Why should the nominalists not simply say, "Yes, we have no account of the reliability of applied mathematics. We accept it as reliable simply because it has always worked."? Section 5 is a defense of the following answer to this question. There is no objection to their saying this, provided they are willing to say that our use of mathematics is like the use of calculators whose principles of operation no one understands and which are trusted only because they always work.

Chapter 4

Reply to Peter van Inwagen's Opening Statement

William Lane Craig

Contents

1. Introduction

Peter van Inwagen's meticulously crafted opening statement is a welcome clarification and development of his views on ontological commitment and goes a considerable distance, I think, toward closing the gap between us. The heart of his essay is to be found in his Sections 5–7 in Chapter 1. But before we look more closely at those sections, it deserves to be mentioned at the outset that van Inwagen has said almost nothing in his Opening Statement by way of answer to the question under debate, namely, *do abstract objects*, in particular numbers, *exist?* His essay focuses almost entirely on the articulation and defense of a neo-Quinean criterion of ontological commitment, which, as we have seen, serves as premiss (I) of some sort of Indispensability Argument for Platonism.[1] But he says

1. As is clear in my opening essay, as well as in *God over All*, the formulation of the sample Indispensability Argument is not mine, as van Inwagen seems to suggest, but Mark Balaguer's, nor is it presented by Balaguer as van Inwagen's considered view. It is a rough schema for capturing the gist of Indispensability Arguments. Van Inwagen rejects the claim that singular terms are, as such, devices of ontological commitment but accepts such a claim for existential quantifiers. That claim will play the role of (I) in his argument.

DOI: 10.4324/9781003008712-6

very little about the various responses to premiss (II) of the Indispensability Argument, such as absolute creationism, conceptualism, arealism, fictionalism, *ultima facie* strategies, and pretense theory.[2] Since these alternatives to Platonism are consistent with van Inwagen's proffered criterion of ontological commitment, he has yet to offer a prima facie case for the reality of abstract objects. Since he will doubtless address these alternatives in his Reply to my Opening Statement (which I have, of course, not yet read), I shall not say anything further about them at this juncture.

2. Van Inwagen's Criterion of Ontological Commitment

What, then, about van Inwagen's formulation and defense of premiss (I)?

2.1 Neutralism and the "Two Kinds Fact"

Van Inwagen responds at length to philosophers, such as neutralists and neo-Meinongians, who do not accept a neo-Quinean criterion of ontological commitment. The neutralist and neo-Meinongian hold that both the informal quantifiers of ordinary language and the formal quantifiers that, on van Inwagen's view, abbreviate such expressions are ontologically neutral. It is important to differentiate this neutralist view from a deceptively similar view called *quantifier*

2. Van Inwagen briefly criticizes absolute creationism on the grounds that (1) it is hard to make sense of God's *deciding* to create a property, and (2) creation is a causal relation, and abstract objects like properties and propositions cannot enter into causal relations (pp. 25–26; cf. Peter van Inwagen, "Did God Create Shapes?" *Philosophia Christi* 17 [2015]: 285–90). With respect to (1), however, absolute creationists do not typically root necessarily existing abstract objects in God's will but in God's nature (Thomas V. Morris and Christopher Menzel, "Absolute Creation," *American Philosophical Quarterly* 23 [1986]: 355–56, 360; Brian Leftow, *God and Necessity* [Oxford: Oxford University Press, 2012]). Hence, though properties and propositions depend ontologically on God, they are not the result of His deciding. As for (2), while *abstracta* have no causal powers and so cannot be causes, why can they not be effects? Philosophers of aesthetics who are, like van Inwagen, realists about fictional characters usually take them to be abstract objects created by their authors (Christy Mag Uidhir, "Introduction: Art, Metaphysics, and the Paradox of Standards," in *Art and Abstract Objects*, ed. Christy Mag Uidhir [Oxford: Oxford University Press, 2012], p. 7). So why could God not be the author of abstract objects?

variance, a view which makes a brief appearance in van Inwagen's original footnote 58, but which I omitted from my taxonomy of responses to the Indispensability Argument because of its perceived implausibility. Van Inwagen sometimes characterizes neutralism as though it were a form of quantifier variance. He says, for example, that "the central thesis" of a neutralist meta-ontology is "that 'there is' *et al.* have both a lightweight and a heavyweight sense" (p. 68). That may describe quantifier variance, but it is a mischaracterization of neutralism. Van Inwagen compounds the confusion by equating the neutral sense—the only sense for the neutralist of such quantificational phrases—with the lightweight sense.[3] Neutralists, he says, postulate "that expressions like 'there is an ___' and '___ exists' and 'there are ___s' and '___s exist' have both a 'heavyweight' or 'existentially loaded' sense and a 'lightweight' or 'ontologically neutral' sense" (p. 58). So "When we say things like 'There's a lack of compassion in the world' and 'There's a good chance that won't happen,' we undertake no ontological commitments because we are using 'there's' in its ontologically neutral sense" (p. 63). No, the neutralist holds that quantificational expressions are simply neutral with respect to ontological commitments and can be used to make either heavyweight or lightweight assertions depending on the speaker's intentions.

So it seems to me that the neutralist can find perfectly congenial van Inwagen's claim that "although *operators* like 'there is' and 'there exist' do not have heavyweight and lightweight senses, and even if *predicates* like 'exists' do not have heavyweight and lightweight senses, *sentences* in which those operators and predicates occur can have heavyweight and lightweight *readings*" (p. 61; cf. p. 59). That claim seems perspicuous on a neutralist view: the existential quantifiers of ordinary language and the formal quantifier that codifies them are just ontologically neutral, but sentences featuring such quantifiers can have a heavyweight or a lightweight meaning (taking *readings* to be something like *interpretations*). As I indicated, there are contextual clues and rhetorical devices by means of which the speaker can make his meaning clear. Similarly, as van Inwagen says, we can discern the speaker's meaning "by using the methods that one normally employs to determine,

3. Admittedly, I myself have been guilty of such a slip up. For example, in my opening statement I say that "Ayer's first sentence is a neutral quantificational claim," when I should have said "lightweight quantificational claim."

in the conversational situation in which one finds oneself, the meaning of a statement that might be interpreted in more than one way—attention to context, common sense, and so on" (p. 57; cf. p. 59).

Such an account seems to give a clear and plausible explanation of what van Inwagen calls "the Two Kinds Fact," namely the fact that sentences of our ordinary discourse that contain words and phrases like "exist" and "there is a" are of two kinds: sentences that convey ontological commitment and sentences that do not. The explanation is that such expressions are in themselves ontologically neutral but can be employed to make heavyweight or lightweight claims according to the speaker's intentions. What could be more perspicuous?

Van Inwagen objects to the neutralist's explanation of the Two Kinds Fact on the ground that the neutralist postulates that quantificational expressions "have both a 'heavyweight' or 'existentially loaded' sense and a 'lightweight' or 'ontologically neutral' sense" (p. 58). But that describes quantifier variance, not neutralism, which denies that the quantifier as such has two meanings. So we see that neutralism and van Inwagen's brand of Platonism are not so far apart as we might have thought. Both seem to agree that the informal quantifiers of ordinary language are in themselves ontologically neutral and can be used to make either metaphysically lightweight or heavyweight assertions.

2.2 Van Inwagen's Explanation of the "Two Kinds Fact"

What, then, of van Inwagen's explanation of the Two Kinds Fact? Van Inwagen's Platonism seems paradoxical: How could sentences featuring quantificational expressions fail to carry ontological commitments if the expressions have the same meaning they do in sentences that do involve such commitments? It is after all in virtue of such quantificational expressions that certain sentences carry ontological commitments for their users, so how or why are such commitments voided for other similarly quantificational sentences?

Van Inwagen's answer to that question is that the surface grammar of lightweight sentences is misleading. Despite appearances, lightweight sentences do not express propositions to the effect that *There exists a* ____, whereas heavyweight sentences do. Despite their syntactical structure, they are not *really* quantificational statements—at least in van Inwagen's so-called Revised English, an artificial language fragment requiring revisions to the quantificational apparatus of ordinary English. When a heavyweight sentence is put into Revised English, the translation will contain the existential quantifier "Something is such that," which van Inwagen assumes to be a device of ontological commitment. But when a lightweight sentence is rendered in Revised English, one must offer some paraphrase of the original English sentence that avoids the existential quantifier of Revised English.[4] By so doing, we obviate the apparent ontological commitments of the original English sentence. The difference between the role of "there is" in a heavyweight sentence and its role in a lightweight sentence is simply that the Revised English translation of the heavyweight sentence has a syntax that more or less "maps" the English "there is" onto the quantifier "something is such that," whereas this is not the case with the lightweight sentence (p. 61).

2.2.1 *The Need of a Paraphrase*

2.2.1.1 The Feasibility of Paraphrase

Now the matter of finding a paraphrase to avoid the apparent ontological commitments of ordinary language turns out to be a very delicate matter, which van Inwagen merely notes in passing. In his footnote 18, he observes that his proffered paraphrase of a sample lightweight sentence is but "*one* proposal about how one might best express the proposition" that it expresses in Revised English. "It is, of course, often a vexed question whether a certain sentence of Language A is a satisfactory translation into A of a certain sentence of Language B" (p. 61).[5] That goes double when the translation of

4. It's not clear to me whether van Inwagen thinks that the paraphrase must be first offered in ordinary English before its translation into Revised English or whether the translation of the original sentence into Revised English consists in providing a paraphrase in Revised English.
5. Van Inwagen elsewhere points out that there are two reasons why there is no such thing as the unique translation of some sentence into the quantifier-variable idiom (or, what amounts to the same thing, Revised English sentences): (i) the quantifier-variable idiom is present in different degrees in various translations of the original sentence, and (ii) there are alternative, creative ways of translating the original sentence into the quantifier-variable idiom (Peter van Inwagen, "Meta-ontology," in *Ontology, Identity, and Modality: Essays in Metaphysics*, ed. Peter van Inwagen [Cambridge: Cambridge University Press, 2001],

an existentially quantified sentence is expected to avoid existential quantification in the paraphrase!

Here van Inwagen faces a two-fold challenge: First, since he wants to avoid the most bizarre commitments of ordinary language, he needs to provide a way of paraphrasing ordinary language sentences which carry unwanted ontological commitments that eliminates those commitments. Second, he needs to show that anti-realist paraphrases cannot similarly be found for ordinary language sentences seemingly involving commitment to abstract objects like properties. But van Inwagen admits that he can do neither.

With respect to the first task he says that he would like to be able to show that it is always possible to provide a paraphrase of sentences about various unwanted objects, but "to do that, I think, it would be necessary to discover a general, universally applicable way of paraphrasing ordinary sentences of the kind we are interested in," which he admits he cannot do.[6] So, for example, for want of an acceptable paraphrase, van Inwagen's neo-Quineanism leads him to the bold, if nuanced, affirmation that fictional characters like Sherlock Holmes exist.[7]

With respect to the second task he says,

> I cannot hope to provide an adequate defense of this position, for an adequate defense of this position would have to take the form of an examination of all possible candidates for nominalistically acceptable paraphrases of such sentences, and I cannot

pp. 23–4). These factors undermine the objectivity of one's ontological commitments, in which case the neo-Quinean criterion is of limited use in revealing one's ontological commitments. Achille Varzi, without challenging the Quinean criterion of ontological commitment, thus attacks its usefulness in disclosing the ontological commitments of ordinary language speakers. For there is no objective procedure for determining the correct logical form of a sentence, and many paraphrases are simply ingenious attempts to avoid reference. If the paraphrase avoids some of the ontological commitments of the original sentence, how shall we decide which statement to take as true? Varzi charges that van Inwagen's linguistic revisionism involves "a plain misconstrual" of the statements of ordinary language speakers (Achille C. Varzi, "Words and Objects," in *Individuals, Essence, and Identity*, ed. Andrea Bottani, Massimiliano Carrara, and Pierdonisk Giaretta, Topoi Library 4 [Dordrecht: Kluwer Academic Publishers, 2002], p. 65).

6. Peter van Inwagen, *Material Beings* (Ithaca, NY: Cornell University Press, 1990), p. 108.
7. Peter van Inwagen, "Quantification and Fictional Discourse," in *Empty Names, Fiction, and the Puzzles of Non-Existence*, ed. Anthony Everett and Thomas Hofweber (Stanford, CA: CSLI Publications, 2000), pp. 235–46.

> hope to do that. . . . My statement "We can't get away with [nominalism]" must be regarded as a promissory note.[8]

But a mere promissory note is plainly inadequate as a defeater of anti-realist views. After all, prior to the work of Chihara and Hellman, alluded to in my opening statement, philosophers had widely despaired of finding paraphrases of mathematical sentences which were not ontologically committing to mathematical objects. But now such paraphrases have been formulated.

2.2.2 The Necessity of Paraphrase

One might further question why one's ontological commitments should hinge so crucially on the availability of a paraphrase in Revised English, especially given the lack of objectivity of the commitments of the paraphrases and the dubiousness of some of those commitments. Van Inwagen stipulates,

> Parties to the dispute who are unwilling to accept some ontological implication of a rendering of some thesis they have affirmed into the quantifier-variable idiom must find some other way of rendering that thesis into the quantifier-variable idiom (must find a paraphrase) that they are willing to accept and which does not have the unwanted implication.[9]

"Must?" Why? It is not as though our sentence is meaningless in the absence of a paraphrase, for it is admitted on all hands that our lightweight sentences have linguistic meaning. We can look for paraphrases if we want to, in hopes of showing our interlocutor that we are making a merely lightweight claim. But if we are not ingenious enough to find acceptable paraphrases for various lightweight sentences of ordinary language—and sometimes very considerable ingenuity, indeed, is required—are we really committed to such bizarre entities as would be required by a heavyweight reading of those sentences, on pain of the falsehood of our discourse?

8. Peter van Inwagen, "A Theory of Properties," in *Oxford Studies in Metaphysics*, ed. Dean Zimmerman (Oxford: Oxford University Press, 2004), pp. 118–19.
9. Peter van Inwagen, "Being, Existence, and Ontological Commitment," in *Metametaphysics: New Essays on the Foundations of Ontology*, ed. David Chalmers, David Manley, and Ryan Wasserman (Oxford: Clarendon, 2009), p. 506.

It might even be the case that English lacks the linguistic resources for anyone to formulate adequate paraphrases of all the sentences we rightly take to be lightweight. Should ontological commitments be foisted upon us by what might be a historically contingent deficit of our native language? If the neo-Quinean demands to know what proposition is expressed by "Numbers exist" in ordinary English, the neutralist can say what van Inwagen himself says with respect to a given lightweight sentence: that he has no better way of expressing it than that sentence taken in a lightweight sense.[10] As indicated in my opening statement, we can illustrate the difference by appealing to examples where paraphrases are available, *e.g.*, the sentence "The number of Martian moons is two" expresses the same proposition in English as "Mars has two moons," which is not ontologically committing to the number two. But there is no reason to expect that English should have the resources to paraphrase away any given quantification over or reference to things, given the lightweight sense of so many existential sentences of ordinary language. It seems to me, then, that the demand for a paraphrase is nothing more than neo-Quinean imperiousness.

2.2.3 The Propositional Content of Paraphrases

Moreover, why think that the propositions expressed by lightweight English sentences are the propositions expressed by acceptable Revised English paraphrases? Some of those paraphrases are nothing short of bizarre. For example, consider the ordinary English sentence, "There are two valuable chairs in the next room," which for van Inwagen is a lightweight claim.[11] Since van Inwagen believes that there are no inanimate composite entities, but (apart from living things) just fundamental particles which are differently arranged, he thinks that chairs do not exist. But van Inwagen is reluctant to say that the beliefs of the average man with respect to chairs and other material objects are therefore false. That would contravene van Inwagen's deeply held belief that most of the things we believe are true.[12] So he is quite sensitive to the objection that

10. Peter van Inwagen, "Introduction: Inside and Outside the Ontology Room," in *Existence: Essays in Ontology* (Cambridge: Cambridge University Press, 2014), pp. 15–16.
11. Van Inwagen, *Material Beings*, p. 107.
12. For an expression of that belief see Peter van Inwagen, "God and Other Uncreated Things," in *Metaphysics and God: Essays in Honor of Eleonore Stump*, ed. Kevin Timpe (London: Routledge, 2009), p. 19.

his metaphysical views about composite objects imply that many of our commonly held beliefs turn out to be false. He asks,

> Is the existence of chairs—or, at any rate, of things suitable for sitting on, like stones and stumps—a matter of Universal Belief? If it were, this would count strongly against my position, for any philosopher who denies what practically *everyone* believes is, so far as I can see, adopting a position according to which the human capacity for knowing the truth about things is radically defective. And why should he think that his own capacities are the exception to the rule?[13]

So van Inwagen essays to prove that his view of material objects "does not contradict our ordinary beliefs."[14] His defense is:

> It is far from obvious, however, that it is a matter of Universal Belief that there are chairs. In fact, to say that any particular proposition that would be of interest to philosophers belongs to the body of Universal Belief is to put forward a philosophical thesis and no trivial one. It is difficult to settle such questions, in part because there are lots of things that one might express by uttering 'philosophical' sentences like 'There are chairs', and some of them might be things that are irrelevant to the concerns of ordinary life. It may be that the intellectual training provided by dealing with ordinary matters ill equips one to appreciate them.[15]

This defense strikes me as inadequate. I think it is indisputable that the vast majority of people believe that there are objects that one can sit on like chairs and stones and stumps. So saying is *not* to put forward a philosophical thesis but a sociological thesis. What is a philosophical thesis is the thesis that there are objects like chairs, and the average person accepts that thesis almost unthinkingly.[16] Certainly there are many things that the neo-Quinean might regard as the propositional content expressed by utterances like "There are

13. Van Inwagen, *Material Beings*, p. 103.
14. Ibid., p. 98.
15. Ibid., p. 103.
16. Van Inwagen seems to think that average people do not hold metaphysical beliefs about the existence of objects, which seems to me clearly wrong. The average person believes that there are objects like chairs, just as he believes that the external world is real or that time is tensed.

chairs" other than that there are chairs, and these are so recondite that it is likely that the common man believes none of them. So, for example, should one say of two chairs in room 103 of the Morris Inn, "There are two very valuable chairs in the next room," then in order to avoid ontological commitment to chairs, van Inwagen must come up with a Revised English paraphrase that discloses the actual propositional content of that lightweight sentence. He proposes,

> There are *x*s and there are *y*s such that [the *x*s are not the *y*s and both the *x*s and the *y*s are arranged chairwise and both the *x*s and the *y*s are in room 103 of the Morris Inn and both the *x*s and the *y*s are collectively very valuable and, for any *z*s, if those *z*s are arranged chairwise and are in room 103 of the Morris Inn and are collectively very valuable, then those zs are the *x*s or those *z*s are the *y*s].[17]

The trick is co-opting what most people take to be the evident truth of the ordinary language expression in order to imply their belief in the Revised English paraphrase. This is illegitimately borrowed capital. When the alleged propositional content of a supposedly lightweight English sentence is revealed by its translation into Revised English, it seems highly implausible that the average person believes the proposition expressed in Revised English.

Here van Inwagen executes a deft move that the reader may have overlooked. In order to render plausible his claim that ordinary language speakers really do believe the proposition expressed by the Revised English paraphrase of a sentence taken to be lightweight, van Inwagen has to assume that one may be mistaken, even upon reflection, about what one really believes. He writes,

> I will assume the truth of the following thesis about belief:
>
> > For any sentences *x* and *y*, if *x* and *y* express the same proposition, the predicates ⌜believes that *x*⌝ and ⌜believes that *y*⌝ are true of exactly the same things.
>
> Thus, if 'All Cretans are liars' and 'All Cretans are mendacious' express the same proposition, then 'believes that all Cretans are liars' is true of Epimenides if and only if 'believes that all Cretans are mendacious' is true of him (pp. 51–52).

17. Personal correspondence, Peter van Inwagen to William Lane Craig.

So if Epimenides believes that "There are two valuable chairs in the next room," then he also believes that "There are *x*s and there are *y*s such that [the *x*s are not the *y*s and both the *x*s and the *y*s are arranged chairwise and both the *x*s and the *y*s are in the next room and both the *x*s and the *y*s are collectively very valuable and, for any *z*s, if those *z*s are arranged chairwise and are in the next room and are collectively very valuable, then those zs are the *x*s or those *z*s are the *y*s]," since these sentences express the same proposition. You ask him, "Epimenides, are there *x*s and are there *y*s such that [the *x*s are not the *y*s and both the *x*s and the *y*s are arranged chairwise and both the *x*s and the *y*s are in the next room and both the *x*s and the *y*s are collectively very valuable and, for any *z*s, if those *z*s are arranged chairwise and are in the next room and are collectively very valuable, then those zs are the *x*s or those *z*s are the *y*s]?" He replies—sincerely—"Certainly not! Nothing could be further from the truth." Nevertheless, on van Inwagen's view, he does believe that there are *x*s and there are *y*s such that the *x*s are not the *y*s and both the *x*s and the *y*s are arranged chairwise and both the *x*s and the *y*s are in the next room and both the *x*s and the *y*s are collectively very valuable and, for any *z*s, if those *z*s are arranged chairwise and are in the next room and are collectively very valuable, then those zs are the *x*s or those *z*s are the *y*s. One might conclude that ordinary language speakers are the least reliable people to ask what they believe; rather we should ask neo-Quinean philosophers.

By his deft move, van Inwagen is able to affirm that someone who in the ordinary business of life asserts that "Shadows exist" has said something true *simplicter*, true *full stop*, true *period,* because the proposition expressed by this sentence is the same as the proposition expressed by the Revised English sentence, "Something is such that it is a shaded portion of some thing's surface," which he accordingly also believes. (Unfortunately, that sentence seems to commit him ontologically to portions and perhaps surfaces, what van Inwagen elsewhere calls arbitrary undetached parts[18]; but never mind.) Van Inwagen is thus able to maintain his personal commitment to the fact that most of what we believe, e.g., that there are chairs, that there is a hole in my shirt, that the Equator bisects the Congo, is true, even if we repudiate the Revised English paraphrases of what we consciously believe.

18. Peter Van Inwagen, "The Doctrine of Arbitrary Undetached Parts," *Pacific Philosophical Quarterly* 62 (1981): 123.

Van Inwagen's assumption strikes me as very implausible.[19] Philosophers who have discussed the nature of belief often provide illustrations which have the effect of falsifying that assumption. For example, John believes that Lewis Carroll wrote *Through the Looking Glass*, but he does not believe that Charles Dodgson wrote *Through the Looking Glass*, since he has no idea that Charles Dodgson was Lewis Carroll. Ancient Babylonian astronomers believed that Phosphorus appears in the morning, but they did not believe that Hesperus appears in the morning, since they were unaware that Phosphorus is Hesperus. If I am lying in hospital suffering from amnesia, I believe that I am lying in hospital, but I do not believe that Bill Craig is lying in hospital, since I do not believe that I am Bill Craig.

Now van Inwagen gives no argument for his pivotal assumption. So I simply ask the reader, which is more plausible: that for any sentences x and y, a person may believe x but not believe y, even though the same proposition is expressed by x and y, or that a person who believes x also believes y, despite his adamant denial of y, if the same proposition is expressed by x and y? If we do not affirm van Inwagen's assumption, it will have reverberations throughout the structure of his argument. For then we shall not find it plausible that a person who holds the lightweight beliefs "There are chairs" or "There are numbers" also believes the relevant, often recondite, paraphrases of Revised English, even if these express the same

19. William Alston lays out what he takes to be "a familiar, ordinary concept of belief, . . . one that is dominant in philosophy" by providing the following list of lawlike conditionals, the truth of which is sufficient to identify the concept of a person's belief that p ("xBp"):

    A. If xBp, then if someone asks x whether p, x will have a tendency to respond in the affirmative.
    B. If xBp, then if x considers whether it is the case that p, x will tend to feel it to be the case that p, with one or another degree of confidence.
    C. If xBp, then x will tend to believe propositions that he takes to follow from p.
    D. If xBp, then x will tend to use p as a premise in theoretical and practical reasoning where this is appropriate.
    E. If xBp, then if x learns that not-p, x will tend to be surprised.
    F. If xBp, then x will tend to act in ways that would be appropriate if it were the case that p, given x's goals, aversions, and other beliefs.

    (William P. Alston, "Audi on Nondoxastic Faith," in *Rationality and the Good: Critical Essays on the Ethics and Epistemology of Robert Audi*, ed. Mark Timmons, John Greco, and Alfred Mele [Oxford: Oxford University Press, 2007], pp. 129, 131). Except for the last, these conditions are incompatible with van Inwagen's assumption.

propositions. A person who holds what he takes to be a lightweight belief that there are chairs or that there are numbers will resist any metaphysically heavyweight Revised English translation of these claims and may not believe any of the proffered Revised English paraphrases. Accordingly, given the prevalence of heavyweight beliefs like "Chairs exist" and lightweight beliefs like "There are odd numbers," it follows that on van Inwagen's view much of what practically everyone believes is false, in contrast to van Inwagen's deeply held belief that most of the things we believe are true.

2.2.4 Ramifications for Van Inwagen's View

What we have said casts doubt, in turn, upon van Inwagen's claim that the propositions expressed by ordinary English sentences are the propositions expressed by acceptable Revised English paraphrases. For if they were, we should be widely mistaken about what propositions are expressed by our assertions and so widely in error about the truth of our beliefs. That renders van Inwagen's explanation of the Two Kinds Fact implausible, for it depends on the assumption that the syntax of lightweight quantificational sentences is systematically misleading when it comes to their propositional content. Their actual propositional content may be the proposition expressed by a Revised English sentence which we do not believe. In note 15, van Inwagen clarifies that he is not claiming that his explanation of the Two Kinds Fact is better than the neutralist's explanation but merely that his explanation "is consistent with my position" (p. 59). Small potatoes, forgive me! If the neo-Quinean's explanation is no better than the neutralist's, then we have no reason to embrace it.

Worse, it follows, in turn, that van Inwagen's criterion of ontological commitment is subverted. For van Inwagen defines a person's ontological commitments in terms of what that person believes there to be (p. 45). Van Inwagen defines ontological commitment using the quantificational expression "something is such that" of Revised English:

> S. *A* is directly ontologically committed to ___s $=_{def.}$ *A* believes that something is such that it is a(n) ____.

where instances of (S) are obtained by replacing "*A*" with a word or phrase that designates a person and replacing "___" with a plural count-noun or nominal phrase.

So if I have the lightweight belief "Some numbers are odd" as expressed in ordinary language but do not have the heavyweight belief "Something is such that it is an odd number" as expressed in Revised English, even if these express the same proposition, then I am not ontologically committed to numbers. I am therefore not required to paraphrase my lightweight belief, since I do not share the belief as expressed in Revised English. Van Inwagen protests that if we give a lightweight reading to "Some numbers are odd", we are saying that we will not regard its syntax as a reliable guide to the syntax of the sentence of Revised English with which it is to be replaced in the ontology room. "My problem is this: I do not know what that 'replacement sentence' could possibly be" (p. 68).[20] We have no idea of what a "nominalistically acceptable paraphrase" would look like. To which I say: So what? I see no need to paraphrase my claim into van Inwagen's strange, artificial language fragment Revised English. When I hear people talking "in the ontology room," I'm inclined to just keep on walking. In short, the neo-Quinean demand for paraphrases in order to avoid ontological commitments is very problematic, indeed, a veritable Achilles Heel of van Inwagen's brand of Platonism.

2.3 "Something Is Such That" and Ontological Commitment

More directly, why should we think that the expression "something is such that" is a device of ontological commitment? It is merely the awkwardness of this expression that accounts for its rarity in ordinary language, whether to make heavyweight or lightweight claims. But in ordinary English we could use it to make lightweight assertions. For example, if we say, "There's a glitch in the plans," and someone replies, "No, there's not," we might answer, "Well, *something* is such that it's a glitch in the plans because they're not working." So why should we assume that in Revised English, this quantificational expression is a device of ontological commitment?

Van Inwagen asserts that since sentences in canonical logical notation involving formal quantifiers and variables are nothing more than abbreviations of certain sentences of Revised English,

20. This is my example, not van Inwagen's. He may have a way of paraphrasing mathematical sentences in terms of properties (p. 7, n. 2), though that still leaves the would-be anti-Platonist with abstract objects.

then all *philosophical* theses about quantification should be understood and evaluated as theses about the sentences of Revised English of which sentences containing formal quantifiers and variables are abbreviations (p. 37). As van Inwagen says, "the *real* existential quantifier is 'something is such that'" (p. 44). So why believe the philosophical thesis that "something is such that" in Revised English is a device of ontological commitment?

Van Inwagen says that the existential quantifier is a device of ontological commitment only in the sense provided by the following statements:

1. The existential quantifier is the phrase "something is such that."
2. A person's direct ontological compendium consists of the propositions that person accepts that are expressed by sentences that are instances of the schema "Something is such that it is a(n) ___".
3. A person's ontological commitments are determined by that person's direct ontological compendium (together with that person's non-existential beliefs) (p. 53).

The problem is that this sense is too thin to support Platonism. For (1) leaves us in the dark as to whether "something is such that" is not just neutral as to ontological commitments. So when it comes to (2), it is not clear whether one's direct ontological compendium cannot include propositions expressed by sentences like "Something is such that it is a crack in the Liberty Bell," "Something is such that it is the most famous detective of English fiction," "Something is such that it is √9," and so on, which are taken to be metaphysically lightweight. Thus, when we come to (3), it is unclear what is the significance of one's ontological commitments.

It seems to me that we need to have "some sort of 'pre-analytic' grasp of the concept of ontological commitment, a grasp that we have independently of and prior to our understanding of the existential quantifier" (p. 43). Fortunately, I think that we do have such a pre-theoretic grasp of the concept. The average man, who has no understanding at all of the existential quantifier, can easily grasp the statement: "I believe that there are, strictly and literally speaking, ____s; that is, I believe that there are such *objects* or *entities* as ____s"—which van Inwagen concedes would be "an equally good way" to express ontological commitment (p. 46). Now if that is the case, then we can reformulate van Inwagen's criterion of direct ontological commitment as

> S*. *A* believes that there are, strictly and literally speaking, ____s; that is, *A* believes that there are such *objects* or *entities* as ____s $=_{def.}$ *A* believes that something is such that it is a(n) ____.

So when van Inwagen adopts the expression "Something is such that it is a(n) ___," it is simply *stipulated* that in Revised English the expression "something is such that" is not ontologically neutral but is a way of stating that there are, strictly and literally speaking, ____s; that is, that there are such *objects* or *entities* as ____s.

In Section 7, van Inwagen addresses the question whether "something is such that" has both a lightweight sense and a heavyweight sense—or, more accurately, is neutral (p. 62). His response is intriguing:

> Craig's reasons for saying that 'exists' and 'there is' have two senses, even if they are valid, do not apply to the phrase 'something is such that'. (Either as a phrase of Revised English or as a phrase of English—for its meaning in Revised English is its meaning in English: it is one of the phrases of English that was 'left just as it was' by the sequence of modifications by which English was turned into Revised English.) (p. 66)

Whoa! This is just the old Quinean claim that the meaning of the existential quantifier is determined by the meaning of the informal quantifiers of ordinary English, only now restricted to "something is such that." That move sits very ill with van Inwagen's recognition that there is no such datum as the Two Kinds Fact "in respect of people's use of 'something is such that'—if only because 'people' do not use it" (p. 67). Since the expression is not used, we are left with agnosticism as to the import of this phrase. So when van Inwagen goes on to say, "It seems evident to me (at any rate) that there is no analogue of the Two Kinds Fact for 'something is such that'" (p. 67), his opinion goes beyond the data of ordinary language. Moreover, since the expression belongs to the English *langue*, if not *parole* (p. 67, n. 23), we are free to use it if we wish, despite its awkwardness, and can use it to assert lightweight sentences, as we saw concerning the glitch in the plans.

Curiously, van Inwagen proceeds to say that "If 'there is' has both lightweight and heavyweight sense [*read*: is neutral], it seems inevitable to suppose that 'something is such that' has a lightweight and a heavyweight sense [*read*: is neutral]" (p. 68). This admission puts "something is such that" in the same boat with the other informal quantifiers of ordinary language. But then it seems mysterious how they, but not "something is such that," can be used to make lightweight claims in ordinary language. For

1. "Something is such that" cannot be used to make lightweight ontological claims in Revised English.
2. "Something is such that" has the same meaning in ordinary English as in Revised English.
3. "Something is such that" has the same meaning as the other informal quantifiers of ordinary English.
4. Therefore, the informal quantifiers of ordinary English cannot be used to make lightweight ontological commitments.

If this correct, van Inwagen does not have a consistent explanation of the Two Kinds Fact after all. On the other hand, if the informal quantifiers of ordinary language are neutral, then so is "something is such that," even in Revised English, which completely undermines the neo-Quinean criterion of ontological commitment.

2.4 The Meaning of "Exists"

So van Inwagen returns to the question of the meaning of ordinary language quantificational expressions. He says that before we can answer what "existentially neutral" means, we need to answer the prior question, "What does 'exist' mean?" (p. 69). A bold venture! Van Inwagen contends that "'exists' can be defined in terms of '~', '=', and 'everything is such that' ['∀']" (p. 70), symbols or words whose meaning he esteems to be relatively uncontroversial:

$$x \text{ exists} =_{def.} \sim(\forall y)\ (\sim [y = x])$$

In other words, x exists if and only if there is something identical to x, or more simply if and only if $x = x$. "'Are you saying that existence is nothing but self-identity, then?' Yes, I am" (p. 74). This account glosses over the complaint of free logicians, discussed briefly in my opening statement (p. 110), that in traditional logic identity statements are just as loaded with existence assumptions as existentially quantified statements.[21] Karel Lambert, a pioneer of free logic, argues convincingly in my opinion that this ontological loading of mere identity statements is absurd. We could not then affirm the truth of van Inwagen's statements "Descartes = Descartes" (p. 73) and "Cicero = Tully" (p. 73), statements which would be no more true than "Descartes = Tully," since Descartes and Cicero no longer exist and so there are no things with which they can be identified. Nor could we affirm the truth of identity statements involving the irreferential singular terms that pervade ordinary language like "the dangers of global warming = the dangers of climate change" or "the touchdown scored at the end of the fourth quarter = Brady's only touchdown of the game." Van Inwagen's rejection of free logic's understanding of identity statements saddles him with "the consequence that our definition of ontological commitment. . . will entail that everyone is ontologically committed to self-identical things" (p. 53, n. 11; cf. p. 73, n. 25). So although van Inwagen concludes that if his account of the meaning of "exists" is correct, then it meaningless to say that "something is such that" has either a heavyweight sense or a lightweight sense (p. 75)—a conclusion consistent with neutralism—it remains the case that on his account true identity statements are heavyweight statements carrying ontological commitments, which renders his account implausible.

2.5 Concluding Reflections on Ontological Commitment

So where are we? Van Inwagen does not seem to have a plausible explanation of the Two Kinds Fact that existentially quantified statements of ordinary language can have either a lightweight or a heavyweight sense. Indeed, given his account of "x exists" it seems incomprehensible how such sentences could have a lightweight sense. For that account is equivalent to Quine's view that x exists $=_{def.} (\exists y)\ (x=y)$, which is to say that not just "something is such that" but all the informal quantificational phrases of ordinary language abbreviated by "$\exists$" are devices of ontological commitment. The fact that even identity statements of the form $x=x$ are

21. See discussion in my *God and Abstract Objects*, Chapter 6.

metaphysically heavyweight statements makes it hard to see how lightweight quantificational statements are possible.

Ironically, the neutralist need have no quarrel with van Inwagen's criterion of direct ontological commitment (or its extrapolation to a general criterion of ontological commitment). After all, on occasion, he will want to make metaphysically heavyweight claims like "There are no mathematical objects" or "There are no properties," and then Revised English may be useful to him in articulating his claims. So he may on occasion stop by the ontology room after all. Given the stipulation that "something is such that" is a device of ontological commitment in Revised English, he will simply deny that he believes that something is such that it is, e.g., a number. If he wants to, he can try to provide a paraphrase of his lightweight beliefs, as *ultima facie* strategists do, or just treat the Revised English translations as false, as fictionalists do. But for the most part, he may just refuse to play the game pushed by the neo-Quinean.

Consider van Inwagen's illustration of Sally, who lacks the belief that something is such that it is a well-ordering relation on the real numbers, but who *accepts* all the axioms that define the properties of the real line and *accepts* the axioms of some "first-order" set theory, including the Axiom of Choice. Van Inwagen infers that "Then she is ontologically committed to the existence of well-ordering relations on the real numbers" (p. 54). Wait a minute! Not so fast! We must not overlook the fact that Sally *accepts*—but is not said to *believe*—the relevant mathematical axioms. The distinction between acceptance and belief, well known in the philosophical literature,[22] plays a crucial role in philosophy of mathematics.[23] Sally may be an arealist like Penelope Maddy or a pretense theorist like Mary Leng, who accept the standard set theoretical axioms but do not believe them, as required by van Inwagen's criterion of direct ontological

22. See, e.g., William P. Alston, "Belief, Acceptance, and Religious Faith," in *Faith, Freedom, and Rationality*, ed. Jeff Jordan and Daniel Howard-Snyder (Lanham, MD: Rowman & Littlefield, 1996), pp. 3–27, who draws upon the work of L. Jonathan Cohen, *An Essay on Belief and Acceptance* (Oxford: Clarendon Press, 1992).

23. See Penelope Maddy, "Believing the Axioms I," *Journal of Symbolic Logic* 53/2 (1988): 481–511; idem, "Believing the Axioms II," *Journal of Symbolic Logic* 53/3 (1988): 736–64; and her later *Defending the Axioms: On the Philosophical Foundations of Set Theory* (Oxford: Oxford University Press, 2011), ix. See further Gideon Rosen and John P. Burgess, "Nominalism Reconsidered," in *The Oxford Handbook of Mathematics and Logic*, ed. Stewart Shapiro, Oxford Handbooks in Philosophy (Oxford: Oxford University Press, 2005), 517.

commitment. She would be perfectly consistent to lack the belief that something is such that it is a well-ordering relation on the real numbers, not merely because she lacks the relevant beliefs about mathematical objects expressed in ordinary English, but even more because the belief about the well-ordering relation is expressed in Revised English with its ontologically committing quantifier. What Sally needs to do is have logically consistent lightweight mathematical beliefs and acceptances, but she is not thereby ontologically committed to any mathematical objects.

Once again, we see that the anti-Platonist and van Inwagen are not really so far apart after all. They can agree that if we take the syntactical structure of lightweight sentences of ordinary language quantifying over abstract objects to be a reliable guide to how they should be translated into the fragment of artificial language known as Revised English, then their truth would commit their user ontologically to abstract objects. But whereas the Platonist regards such Revised English sentences as true, the anti-Platonist will regard them as false. This is of little significance, since nobody really speaks Revised English, and the lightweight sentences of everyday language in which we do speak and write that quantify over or contain singular terms referring to abstract objects are agreed by all parties to be true. What we're still wanting is any reason to become a Platonist.

3. God Enters the Picture

As a fellow Christian philosopher, van Inwagen is clearly exercised by my claim that Platonism is theologically unacceptable. The justification for my claim is that according to Christian theology, God is the **sole uncreated reality**. That is not to say that someone who disagrees with that claim is a heretic. The reader might have unfortunately gotten the impression from van Inwagen's reaction that I am some sort of heresy hunter who had accused van Inwagen of heresy. I have neither made nor do I believe such a charge. There are theological mistakes that do not rise to the level of heresy, lest we all be counted heretics! Rather as a Christian philosopher myself, I wanted to determine what theological guidelines there might be concerning the debate over Platonism.

What I discovered surprised me. That God is the sole uncreated reality is the testimony of both the Bible and the ecumenical creeds. The biblical evidence is even of some philosophical interest. What one discovers is that certain didactic passages in the New Testament, particularly in the Fourth Gospel and the letters of Paul, have their background in Middle Platonism, such as was exemplified by the Jewish philosopher Philo of Alexandria. The principal innovation of the Middle Platonists was to move the Platonic realm of the Ideas into the mind of God, sometimes denominated the Logos. This move was congenial to Jewish thinkers like Philo, for the position of Second Temple Judaism (the Judaism that prevailed in Israel up through the time of Jesus) was that everything other than God had been created by God. Neither Hellenistic Jewish thinkers like Philo nor the New Testament authors could countenance an uncreated realm of things, even abstract objects.

I was puzzled when van Inwagen reported that in reciting the Nicene Creed, which affirms that God is the creator of "all things visible and invisible," he adds a silent codicil "that stand in causal relations." Is that what the Nicene Creed requires? Is that a legitimate interpretation of the credal confession? What did the ante-Nicene Church Fathers actually say about God's being the sole uncreated reality? Upon investigation, I must say that I was taken aback at how explicit and forceful they were in addressing this question. The Greek word that played such a crucial role in the debates leading up to the Council of Nicaea was *agenētos*, which means *uncreated*. The ante-Nicene Fathers were unanimously committed to the fact that there is but one *agenētos*, and that is God. They even spoke to the existence of mathematical objects, properties, and forms and denied that they are *agenēta*. Sometimes they underlined the point by imagining God (with His immanent Logos) existing alone prior to creation and affirming that literally nothing else existed. Given this background to the Nicene Creed, it is virtually undeniable that the Creed means to affirm God as the Creator of everything outside Himself, the sole uncreated reality.

Van Inwagen's attempt to avoid this conclusion is unavailing. He has yet to come to grips with the biblical data. He rightly points out that in the ancient world putative *abstracta* stood in causal relations to the physical world and that the Church Fathers did not have any awareness of van Inwagen's peculiar theory of properties. These two facts are red herrings. Possession of causal powers was just an irrelevancy in these debates. There are no grounds whatsoever for thinking that if someone were to hold that numbers and properties and so on are causally effete but nonetheless just as real as elementary particles, indeed, as God Himself, then it would have been theologically acceptable to hold that they exist independent of God. Such acausal *agenēta* would have been excluded along with causally related *agenēta*. There just were no *agenēta* apart from God Himself.

It needs to be appreciated in this connection that van Inwagen is, in spite of his claims, a heavyweight Platonist in the sense that he thinks that abstract objects are just as real as anything else. It is without ontological significance that on his view abstract objects are essentially causally effete (p. 18). What matters is that unlike truly lightweight Platonists, such as Michael Dummett, Bob Hale, and Crispin Wright, who deny that *abstracta* exist in the same robust sense that indisputably existent objects exist, van Inwagen maintains that the predicate "exist" is univocal and that therefore abstract objects are just as real as anything else that truly exists. Abstract objects are for van Inwagen as real as elementary particles, people, and God Himself. But he thinks that they, like God, are *agenēta*.

I agree, of course, with van Inwagen that a resuscitated Church Father would not have any idea at all of what van Inwagen meant by "property" or any inkling of van Inwagen's reasons for thinking that there are such things (p. 9). But what the resuscitated Church Father would and did understand is that God is the sole *agenētos* and that everything else, whatever its nature, no matter how peculiar, if it exists in the same sense that you and I exist, is created by God. Therefore, van Inwagen is just fooling himself if he thinks that after thoroughly explaining his view to the Church Father, he would say anything else than, "'Ah, now I see what you mean by "property". It's wholly unlike anything I had ever thought of. But you're still a heretic if you say they're uncreated.' (Or other words to that effect.)" (p. 10).

Fortunately, there's just no necessity of making van Inwagen's theological compromise. Even without abandoning his favored criterion of ontological commitment van Inwagen has plenty of options open to him other than Platonism. Given his rejection of **constituent ontology**, absolute creationism should be a congenial option for him that would involve very little readjustment. Or he could embrace fictionalism, just as he has embraced an essentially fictionalist view of inanimate composite objects like chairs. Or he could adopt some *ultima facie* strategy, such as figuralism, a very plausible way of avoiding ontological commitments to *abstracta* right in line with his neo-Quinean criterion. Or how about pretense theory, which is a very defensible view that is consistent with his meta-ontology? Indeed, rather than fly in the face of biblical teaching and credal confession, why not instead just renounce the demand for an acceptable paraphrase of lightweight statements quantifying over abstract objects? Such a little step for someone who admits that he cannot paraphrase away all the other unwanted, apparent commitments of ordinary language anyway! Van Inwagen agrees with the neutralist that ordinary language is far from universally heavyweight, so all he need do to retain his criterion is to reject the necessity of a paraphrase. A small price to pay, indeed, to maintain God's status as the sole uncreated reality!

4. Conclusion

In conclusion, van Inwagen still owes us a prima facie case for the reality of abstract objects. Moreover, his case for his criterion of ontological commitment is fraught with difficulty, especially in view of its requirement of acceptable translations of lightweight quantificational English sentences into Revised English. Nevertheless, van Inwagen has succeeded, I think, in narrowing the gap between Platonist and anti-Platonist views of mathematical objects by agreeing that ordinary, true, quantified English sentences have lightweight and heavyweight readings and by providing a language fragment in which, given its stipulated meanings, the anti-Platonist can express his rejection of heavyweight claims about mathematical objects.

Second Round of Replies

Chapter 5

Response to William Lane Craig's Reply

Peter van Inwagen

Contents

In this, the final part of my contribution to this book, I will respond to some criticisms of my views put forward in various passages in Craig's opening statement and his reply to my opening statement. I have been selective; if I attempted to reply to all Craig's criticisms, I should still be writing on the day of my death, and my replies would fill several volumes.

1. Ordinary English, Revised English, and the Ontology Room

What we have said casts doubt . . . upon van Inwagen's claim that the propositions expressed by ordinary English sentences are the propositions expressed by acceptable Revised English paraphrases. For if they were, we should be widely mistaken about what propositions are expressed by our assertions and so widely in error about the truth of our beliefs. That renders van Inwagen's explanation of the Two Kinds Fact implausible, for it depends on the assumption that the syntax of lightweight quantificational sentences is systematically misleading when it comes to their propositional content (p. 189).

DOI: 10.4324/9781003008712-8

It is in fact *not* my position that when, e.g., Theodora the theologian writes,

1. There is but one living and true God, everlasting, without body, parts, or passions,

her sentence expresses the same proposition as the sentences

2. Something^{x} is such that (it^{x} is a living and true God, everlasting, without body, parts, or passions and everything^{y} is such that (if it^{y} is a living and true God, everlasting, without body, parts, or passions, then it^{y} = it^{x}))

and

2′. $\exists x$ (x is a living and true God, everlasting, without body, parts, or passions and $\forall y$ (if y is a living and true God, everlasting, without body, parts, or passions, then $y = x$)).

(The sentences (2) and (2′) must express the same proposition, owning simply to the fact that the latter is an abbreviation of the former.)

Nor is it my position that when Theodora writes

> There is but one path to righteousness, and that is faith in Jesus Christ,

that sentence expresses the same proposition as the sentence

> $\forall x$ (x is human and it is not the case that x is righteous, then x will become righteous if and only if x has faith in Jesus Christ).

I do, however, maintain that the propositions expressed by (1) and (2) are very closely connected—as closely as the propositions expressed by the two sentences:

> The present king of France is bald.
>
> At least one thing is male and now reigns over France; at most one thing is male and now reigns over France; anything that is male and now reigns over France is bald.

And it is true that I have said that if anyone spoke sentence (1) *in the Ontology Room*, that person would be understood by the other inhabitants of the Ontology Room as having affirmed the proposition that $\exists x$ (x is a living and true God, everlasting, without body, parts, or passions and $\forall y$ (if y is a living and true God, everlasting, without body, parts, or passions, then $y = x$)). But that is only because a commitment to so understanding a sentence that starts with 'there is' is one of the defining characteristics of the Ontology Room. It does not imply that that proposition is the proposition Theodora affirmed when she wrote sentence (1). She was not, after all, writing in the Ontology Room.)

"But then what proposition *did* Theodora affirm when she wrote the sentence (1)?"

Why, the proposition that there is but one living and true God, everlasting, without body, parts, or passions. I mean—you speak English, don't you? (If you understand a sentence, don't you know what proposition it expresses?)

"Well, what about Theodora's sentence 'There is but one path to righteousness, and that is faith in Jesus Christ'? What proposition does that sentence express, and how is it related to the proposition expressed by the sentence '$\forall x$ (x is human and it is not the case that x is righteous, then x will become righteous if and only if x has faith in Jesus Christ)'?"

It expresses the proposition that there is but one path to righteousness, and that is faith in Jesus Christ. That proposition is (I should think) not the proposition that $\forall x$ (x is human and it is not the case that x is righteous, then x will become righteous if and only if x has faith in Jesus Christ). But the propositions are pretty intimately related, as intimately as—let's vary the example—the propositions expressed by the sentences

Marion was born in either Idaho or Montana

and

If Marion was not born in Montana, she was born in Idaho.

Or—a perhaps more closely analogous pair of sentences—

Freedom of speech does not exist in North Korea

and

It is not the case that $\exists x$ (x is a North Korean citizen and x enjoys freedom of speech).

And, of course, one would be well advised not to use any sentence that begins 'There is but one path to righteousness' in the Ontology Room or one *will* be confronted with demands for a careful statement of one's ontology of "paths." But if one says (in the Ontology Room), "Any human being who is not righteous can become righteous only by having faith in Jesus Christ" one will not cause oneself to be subjected to demands for answers to bizarre metaphysical questions.

Summary

The proposition that a sentence of ordinary English expresses if it is used in the Ontology Room is not in general the proposition that that sentence expresses if it is used outside the Ontology Room. The proposition that a sentence of ordinary English expresses if it is used inside the Ontology Room is not the "real meaning" of the English sentence.

2. 'Something Is Such That' As a Device of Ontological Commitment

[W]hy should we think that the expression "something is such that" is a device of ontological commitment? It is merely the awkwardness of this expression that accounts for its rarity in ordinary language, whether to make heavyweight or lightweight claims. But in ordinary English, we could use it to make lightweight assertions. For example, if we say, "There's a glitch in the plans," and someone replies, "No, there's not," we might answer, "Well, *something* is such that it's a glitch in the plans because they're not working." So why should we assume that in Revised English, this quantificational expression is a device of ontological commitment? (p. 190)

So when van Inwagen adopts the expression "Something is such that it is a(n) ____," it is simply *stipulated* that in Revised English the expression "something is such that" is not ontologically neutral but is a way of stating that there are, strictly and literally speaking, ____s; that is, that there are such *objects* or *entities* as ____s. (p. 192)

Well, Craig certainly believes that the expression 'there are, strictly and literally speaking, such *objects* or *entities* as ____s' is not ontologically neutral but is a way of stating that there are, strictly and literally speaking, ____s; that is, that there are such *objects* or *entities* as ____s.[1] And I would agree. My position is that 'something is such that it is a(n)'*also* has that property—for, in my view,

> There are, strictly and literally speaking, such *objects* or *entities* as ____s

is nothing more than a long-winded way of saying what

> Something is such that it is a ____

says—as clearly (more so, if anything), as effectively (at least), and more succinctly (certainly).

Craig supposes that if one says, "There are, strictly and literally speaking, universals *ante res*; that is, that there are such *objects* or *entities* as universals *ante res*," one is thereby ontologically committed to universals *ante res*—and without stipulating that 'There are, strictly and literally speaking, ____s; that is, that there are such *objects* or *entities* as ____s' is a device of ontological commitment. And he supposes that if one says, "Something is such that it is an universal *ante res*,"[2] one is not ontologically committed to universals *ante res*, or not unless one has at some earlier moment stipulated that 'something is such that it is a(n) ____' is a device of ontological commitment. Why is it that the latter phrase is a device of ontological commitment only if one stipulates that it is, and the former is a device of ontological commitment without any help from stipulation?

If I interpret Craig correctly, the "glitch in the plans" example is central to his answer to this question. He contends that if Alice says (in circumstances of the sort he imagines: ". . . we might answer . . . .")

1. Craig has said, "...the neutralist holds that quantificational expressions are simply neutral with respect to ontological commitments and can be used to make either heavyweight or lightweight assertions depending on the speaker's intentions." But 'There are, strictly and literally speaking, such *objects* or *entities* as . . .' is a quantificational expression. So what he really believes is that *some* quantification expressions are neutral with respect to ontological commitments and some aren't. The question is, where—and how—does he draw the line?
2. Or simply 'Something is a universal *ante res*'. The function of 'is such that' is to allow quantifiers to be *sentence* operators—as in, for example, 'Something is such that (if it is Greek, then it is mortal)'. But in cases in which the sentence to which 'something is such that' is applied is a very simple one like 'it is a universal *ante res*', 'is such that' has no function and might as well be omitted.

> Something is such that it's a glitch in the plans,

and Bertram says,

> There is, strictly and literally speaking, an *object* or *entity* that is a glitch in the plans,

Alice undertakes thereby no ontological commitments, and Bertram does. (We are supposing that neither has stipulated anything.) For my part, however, I have a hard time seeing any difference between 'something is such that it is a(n)' and 'There is, strictly and literally speaking, an *object* or *entity* that is a(n)'. If something is such that it is a glitch in the plans, how could there not be, strictly and literally speaking, an *object* or *entity* that was a glitch in the plans? Is there something that is not an object or an entity? What would it *be*? What would a thing that was not an object or an entity be *like*?

I suppose one might say, "Well, a glitch is just a flaw, and if one says 'There's a flaw in our plans,' that's really only to say that our plans are flawed[3]—which doesn't imply the existence of anything but us and our plans." But that's really more my line of country than Craig's. That's a *philosopher's invented paraphrase* of 'There a flaw in our plans', not one of two alternative possible meanings the English language provides for that sentence, and which are automatically "there" for a speaker to choose between.

Perhaps, however, the choice of example ("a glitch in the plans") is an avoidable complication. If you asked me what something that was a glitch in some given set of plans would be like, I'd have a hard time answering. Let's have a simpler, more straightforward example:

> Carla says, "Something is such that it's a hole in this piece of cheese," and Derek says, "There is, strictly and literally speaking, an *object* or *entity* that is a hole in this piece of cheese."

What has Derek said that Carla has not said?

---

3. Well, it *might* be really to say only that. But if it makes sense for someone to respond to an utterance of "There's a flaw in our plans" by saying, "Indeed there is. In fact, there are two flaws, one of which is also a flaw in Carla's plan, and the other of which is also a flaw in Derek's plan," then "There's a flaw in our plans" cannot simply mean "Our plans are flawed."

One thing it occurs to me to suppose that Derek's statement might add to Carla's is the following (and in fact it is the only way I can think of).

> Carla's statement might mean only this: that the piece of cheese was shaped in a certain way—that its surface was not (as the topologists say) a genus-0 surface. *That* could be true even if nothing existed but the piece of cheese. Derek's statement, however, is true only if something is such that it's true of it that it's a hole *and* true of it that it's in the piece of cheese.

I myself don't think it's true that Carla's statement could mean only that the piece of cheese was shaped in a certain way. (I think it implies that there's something of which it's true that it's a hole and is in the piece of cheese. But let's suppose I'm wrong and Carla's statement could mean only that the piece of cheese had a certain kind of shape.) If I were convinced that that was the case, I'd simply modify my definitions of the quantifiers; I'd change the sentence in the opening statement (p. 36)

—the word 'something' occurs only in the phrase 'something is such that'. We call this phrase **the specific quantifier.**

to

—the word 'something' occurs only in the phrase 'something is such that *it's true of it that*'. We call this phrase **the specific quantifier.**

Would Craig be willing to make the following statements?

> But in ordinary English we could use 'something is such that it's true of it that' to make lightweight assertions. For example, if we say, "There's a glitch in the plans," and someone replies, "No, there's not," we might answer, "Well, something is such that it's true of it that it's a glitch in the plans because they're not working." (And, of course, saying, "Something is such that it's true of it that it's a glitch in the plans" does not commit one to there being an object or entity that is a glitch in the plans.)

If he did say make those statements, I would disagree. It seems to me that if someone says, "Something is such that it's true of it that it's a glitch in the plans, and someone else says, "There is, strictly and literally speaking, an *object* or *entity* that is a glitch in the plans," then (at least if their two assertions have any meaning at all), the two of them have said the same thing in different words.

Of course, the position Craig is really interested in defending is this: that the existential quantifier of formal logic—the symbol '∃'—is not what he calls a device of ontological commitment. And he will therefore say that any natural-language phrase that I may use to define '∃' is either not a device of ontological commitment or is not what '∃' means. If I say that '∃*x* *x* is an irrational number' means 'Something is such that it is an irrational number', then, since 'something is such that' is not a device of ontological commitment, '∃' (so defined) is not a device of ontological commitment. But if I went the whole hog, if I went so far as to say that '∃*x* *x* is an irrational number' meant 'There is, strictly and literally speaking, an *object* or *entity* that is an irrational number' then '∃' (so defined) would be a device of ontological commitment, but it would not have the meaning it has in logic textbooks. For that reason, I'll say just a bit about the meaning '∃' has in the logic textbooks. I contend that part of what is taught in the logic texts is this

> Any sentence that consists of '∃' followed by a variable followed by a sentence *S* in which that variable alone is free is true just in the case that at least one object satisfies the condition specified by *S*.

So, for example, '∃*x* *x* is an irrational number' is true just in the case that at least one object satisfies the condition specified by '*x* is an irrational number'. And '∃*x* *x* is a glitch in our plans' is true just in the case that at least one object satisfies the condition expressed by '*x* is a glitch in our plans'. It is presumably not the case that at least one object satisfies the condition specified by '*x* is a glitch in our plans'. And it is therefore not the case that '∃*x* *x* is a glitch in our plans' is true. (Meaningless, I'd say—since '*x* is a glitch in our plans' fails to "specify" a condition that an object of some sort might or might not satisfy.) It's my position that 'Something is such that it's a glitch in our plans' is like '∃*x* *x* is a glitch in our plans' in this respect. Craig thinks—on the contrary—that 'Something is such that it's a glitch in our plans' could be used to say something true. If he's right, my definition of '∃' is wrong. Well, if that's so I'll need another definition. I've suggested this possible alteration: '∃*x* F*x*' means 'Something is such that it's true of it that it is F'. But, if necessary, I could go with: '∃*x* F*x*' means 'There is, strictly and literally speaking, an *object* or *entity* that is F'.

Summary

Craig has denied that 'something is such that it is a/(n) ___' is a "device of ontological commitment." In Section 2, it is pointed out that he obviously so regards 'there are, strictly and literally speaking, such *objects* or *entities* as ____s', and it is asked where he draws the line between phrases that are and phrases that are not devices of ontological commitment. It is contended that, wherever the line is to be drawn, 'something is such that it is true of it that it is an *x*' should fall on the "device of ontological commitment" side of it. It is pointed out that van Inwagen might have used 'something is such that it is true of it that it is a/(n) ___' rather than 'something is such that it is a(an)___' in his definition of the existential quantifier.

3. The Things Most People Believe

I think it is indisputable that the vast majority of people believe that there are objects that one can sit on like chairs and stones and stumps (p. 185).

I suppose Craig regards that thesis as indisputable because if you ask "the man on the Clapham omnibus" the question, "Do you believe that there are chairs?", he will reply "Yes."[4] Moreover, he and his fellow passengers on the Clapham omnibus, although they rarely if ever make highly general statements like "There are chairs" or "Chairs exist", do often say things that most philosophers would say logically imply that there are chairs or that chairs exist—"If we need more chairs, there are lots of them in the ballroom," and so on.

But let's look at a case in which it's plausible to suppose that someone does actually use the very general sentence "Chairs exist"—a case I borrow from my "Inside and Outside the Ontology Room":[5]

> "You and I may be brothers, but no two people could be less alike. I have devoted my life to working for peace and justice, and *your* only goal in life is to get rich selling furniture."

4. Actually, I think it more probable that he'd back warily away from you, glancing anxiously about for the safest way off the bus.
5. Compare the discussion of 'Shadows exist' and "the Hopkinsonian proposition" in Section 4 of my Opening Statement.

> "What can I say? I deal in reality and you deal in dreams. Chairs exist. Peace and justice don't and never will."

Consider the proposition the second speaker, the hard-headed, cynical businessman, affirmed when he uttered the words 'Chairs exist'. In my view—for the sake of simplicity, I will pretend that all human beings are familiar with chairs—everyone accepts this proposition. I certainly do. In my view, however—and few philosophers agree with me on this point—that proposition and the proposition that something is such that (it is true of it that) it is a chair are distinct propositions.

Let us give the proper name 'Proposition A' to the proposition—whatever proposition it may have been—that the businessman affirmed when he spoke the words 'Chairs exist'. And let us give the proper name 'Proposition B' to the proposition that something is such that (it is true of it that) it is a chair. Most metaphysicians who have considered the question say this:

> Proposition A is identical with Proposition B. That proposition is true.

One metaphysician (Professor Trenton Merricks) says this:

> Proposition A is identical with Proposition B. That proposition is false.[6]

And I myself—no less lonely than Merricks—say this:

> Proposition A and Proposition B are distinct propositions. Proposition A is true and Proposition B is false.

The phrase 'in the Ontology Room' is so defined that, if the sentence 'Chairs exist' is spoken in the course of a discussion of some ontological question, that discussion takes place in the Ontology room only if all the participants in the conversation agree to understand that sentence as an informal stylistic variant on 'something is such that (it's true of it that) it's a chair'. If, therefore, I am right to suppose that the cynical businessman's utterance of 'Chairs exist' did not express the proposition that something is such that (it's true of it that) it's a chair', 'Chairs exist' expresses different propositions inside and outside the Ontology Room.

6. Merricks (2007, pp. 162–70).

"But why should I—when I'm discussing ontology—pay any attention to those rules you've said define 'the Ontology Room'? Who are you to tell me what I have to mean by sentences like 'Chairs exist'?"

There is no reason whatever for you to pay attention to those conventions. The only consequence of your refusing to adopt them in discussions of ontological questions is that I will decline to discuss ontology with you—for, from my point of view, to discuss ontology in language that is not governed by those rules is a waste of time. Discuss ontology without entering the Ontology Room all you like, but you won't be discussing it with *me*. (But I'll be happy to discuss meta-ontological questions with you—such as the question why I suppose it's a waste of time to discuss ontological questions in language that is not governed by those rules.)

I will now give a more detailed statement of the features I ascribe to propositions A and B.

Some features of Proposition A (according to van Inwagen)

It is necessarily equivalent to (but is not the same proposition as) the proposition that some things are arranged chairwise (that is, are arranged spatially and fastened together as the parts of a chair are if there are chairs).

It is the referent of the phrase 'the proposition that chairs exist' when that phrase is used outside the Ontology Room.

It is *not* the referent of the phrase 'the proposition that chairs exist' when that phrase is used inside the Ontology Room.

It is entailed by Proposition B, but does not entail Proposition B. (Note that this "feature" does not follow from of the definition of the Ontology Room. It is simply van Inwagen's position. The thesis that Proposition A and Proposition B are identical is a thesis that can be *debated* in the Ontology Room.)

It is indisputably true—*not* "true in the loose and popular sense but not true in the strict and philosophical sense"; rather, it's *just plain* true, it's true *without qualification*, it's true *period*, it's true *full stop*.[7]

7. Do not fall into the trap of supposing that, according to my view, a *proposition* might be true outside the Ontology Room and false inside it. What *is* possible is

Used outside the Ontology Room, the predicate '____ believes that chairs exist' is true of a person if and only if that person accepts Proposition A.[8]

Some features of Proposition B (according to van Inwagen)

It is, by definition, the proposition that something is such that (it's true of it that) it is a chair. It is therefore the referent of 'the proposition that something is such that (it's true of it that) it is a chair' whether that phrase is used inside or outside the Ontology Room.

It entails, but is not entailed by, Proposition A.

It is the referent of the phrase 'the proposition that chairs exist' when that phrase is used inside the Ontology Room (owing simply to the constitutive rules of the Ontology Room).

It is *not* the referent of the phrase 'the proposition that chairs exist' when that phrase is used outside the Ontology Room

It is false—but not indisputably so. (In fact, most philosophers regard it as true.) That it is false is a substantive philosophical position.

Used inside the Ontology Room, the predicate '____ believes that chairs exist' is true of a person if and only if that person accepts Proposition B.[9]

---

that a sentence should express a certain proposition *x* outside the Ontology Room and a different proposition *y* inside the Ontology Room. And in some such cases, *x* will be true proposition and *y* a false proposition.

8. Craig thinks there's an important distinction between belief and acceptance. As far as I can see, this distinction is an artifact of the appropriation by various philosophers of 'accept' as a term of art. Consider this bizarre exchange. *A*: "Obama was born in Hawaii." *B*: "Yes, I accept that." *Later: B*: "Obama was born in Kenya." *A*: "Wait—a moment ago you agreed that Obama was born in Hawaii." *B*: "You said that he was born in Hawaii, and I accepted what you said. But I don't *believe* he was born in Hawaii." Note that Roderick Chisholm has offered the following schematic definition of belief: '*S* believes that *p* =Df *S* accepts the proposition that *p*' (Chisholm, 1976, p. 161).
9. Suppose someone asks in the Ontology Room whether President Biden believes that chairs exist. Are they asking whether Biden accepts the proposition expressed by 'Chairs exist' inside the Ontology Room or are they asking whether Biden

So then, is it (from my point of view) the case that the vast majority of people believe that chairs exist? A hard question—hard because (so I say) the proposition expressed by the sentence 'Chairs exist' can vary with context. If the question, "Is it the case that the vast majority of people believe that chairs exist?", is asked outside the Ontology Room, the answer is Yes according to everyone. If the question (same words, but different question) "Is it the case that the vast majority of people believe that chairs exist?", is asked *inside* the Ontology Room, the constitutive rules of the Ontology Room force it to mean, "Is it the case that the vast majority of people believe that something is such that it is a chair?" Most philosophers think the answer to that question is Yes. (For that matter, I suppose that most philosophers think that the vast majority of people believe that that there are, strictly and literally speaking, such *objects* or *entities* as chairs.) My position, and a very lonely position it is, is that almost nobody accepts that proposition (or rejects it). I believe that very few people have ever considered or entertained it. And I believe that the almost universal exasperation that philosophers express when I affirm that thesis is grounded in their failure to distinguish propositions A and B.

In Sections 4–6, I turn from logic, semantics, and ontology to theology. I consider some of Craig's criticisms of my views on the relation between platonism and Christian theology.

Summary

Section 3 is devoted to the question whether Craig is right to say that van Inwagen denies the obvious truth that most people believe that chairs exist. It is contended that the sentence 'Chairs exist' can express more than one proposition. One of them, the proposition expressed by 'Chairs exist' outside the Ontology Room, is indeed accepted by most people; another, the proposition expressed by 'Chairs exist' inside the Ontology Room, is neither accepted *or* rejected by any very great number of people. It is indeed true that most metaphysicians accept it, but very few people are metaphysicians.

accepts the proposition expressed by 'Chairs exist' outside the Ontology Room (the President, let us suppose, has never foot in the ontology room)? In designing the Ontology Room, I chose the former option.

4. The Nicene Creed, the Fathers, and Uncreated Things

I was puzzled when van Inwagen reported that in reciting the Nicene Creed, which affirms that God is the creator of "all things visible and invisible," he adds a silent codicil "that stand in causal relations." Is that what the Nicene Creed requires? Is that a legitimate interpretation of the credal confession? What did the ante-Nicene Church Fathers actually say about God's being the sole uncreated reality? Upon investigation, I must say that I was taken aback at how explicit and forceful they were in addressing this question. The Greek word that played such a crucial role in the debates leading up to the Council of Nicaea was *agenētos*, which means *uncreated*. The ante-Nicene Fathers were unanimously committed to the fact that there is but one *agenētos*, and that is God. They even spoke to the existence of mathematical objects, properties, and forms and denied that they are *agenēta*. Sometimes they underlined the point by imagining God (with His immanent Logos) existing alone prior to creation and affirming that literally nothing else existed. Given this background to the Nicene Creed, it is virtually undeniable that the Creed means to affirm God as the Creator of everything outside Himself, the sole uncreated reality (p. 197).

I didn't report any such thing. Nor did I "report" that, when I'm reciting the words

> We believe in one God,
> the Father, the Almighty,
> maker of heaven and earth,
> of all that is, seen and unseen . . .,

as I'm saying 'the Almighty', I add a silent codicil along the lines of "although there are many things he is unable to do, such as to create a spherical cube, change the past, break promises he has made, and bring about his own non-existence." I don't know whether reciting the Nicene Creed is a part of Craig's communal devotional life, but if it is, I doubt whether, in saying the words "all things visible and invisible,"[10] he adds a silent codicil "except himself, and except for privations and defects." I doubt whether the shopkeeper who says, "What a day! We've sold everything in the shop!" adds a silent codicil, "except the cash register and the counter and the shop cat."

10. 'Of all that is, seen and unseen' and 'of all things visible and invisible' are two widespread translations of 'visibilium omnium et invisibilium'.

But let us turn to a more interesting topic—to Craig's theses about implications of the teaching of the Church Fathers concerning Platonism. Let's revisit a certain fictional member of that august company, the Father who played a role in my Opening Statement—the one who had slept for 1700 years or more. He has awakened, and has learned English, but, at the moment we revisit him, he has learned as little about the theological and metaphysical thinking that went on while he slept as is consistent with his having learned a modern language. He is shown the following text (it is made up of phrases taken from Craig's statements descripting the theology of creation of the Church Fathers), and is asked whether he will assent to its content:

> God is the sole uncreated reality. Everything other than God has been created by God. There exists no realm of uncreated things.[11] There is but one *agenētos*, and that is God. If mathematical objects, properties, and forms exist, they are not *agenēta*. Everything besides God, whatever its nature, no matter how peculiar, if it exists in the same sense as the sense in which you and I exist, is created by God.

And he of course says, "Yes." When he says "Yes" (or "I assent" or "That's all perfectly true" or whatever he says), he affirms a certain proposition. And, I maintain, he is right to affirm that proposition—given the possibilities he is aware of.

But might it be that he would no longer affirm that proposition (in its full generality) if he became aware of possibilities he is unaware of?

Craig is quite sure the Father would continue to say exactly the same things, no matter what possibilities were brought to his attention (pp. 198–199). As far as I am able to determine, his argument in defense of this position is logically the same as the argument employed by Wyman in the following exchange:

WYMAN: Against Dismas of Ancyra, Damian of Cyrene maintained *vehemently* that all legally binding civil marriages

11. I omit at this point Craig's words 'even abstract objects'. Our awakened Father will not understand the words 'abstract object'—a coinage of Quine's—if indeed he has "learned as little about the theological and metaphysical thinking that went on while he slept as is consistent with his having learned a modern language."

are sacramentally valid. And he was quite insistent about that 'all'. He used phrases like 'without any possibility of exception'.

McX: I bet he wouldn't have been so insistent about 'without any possibility of exception' if he'd realized that one of the possibilities was same-sex marriage.

Wyman: Well, I agree that that's not a possibility he would have thought of. But he said 'without any possibility of exception,' not 'without any possibility of exception that *I* can think of'. So *of course* if some contemporary had said to him, "But suppose one of the possibilities was a civil marriage between two men or two women," he'd have said, "Even in that case."

Summary

In Section 4, it is first asserted that (contrary to what Craig has said) that Christians who believe that things that can be said of things are uncreated need not have some unspoken reservation in mind when they affirm in the company of others that God is creator of all things, seen and unseen. It is further asserted that Craig has not presented a satisfactory defense of his position that the Church Fathers would insist that things that can be said of things must (If they exist at all) be created things. His defense of this position is unsatisfactory because it either does not take seriously the possibility that such objects were conceptually inaccessible to the Fathers or discounts its relevance.

5. "Just as Real as Anything Else"

It needs to be appreciated in this connection that van Inwagen is, in spite of his claims, a heavyweight Platonist in the sense that he thinks that abstract objects are just as real as anything else. It is without ontological significance that on his view abstract objects are essentially causally effete. ... What matters is that unlike truly lightweight Platonists, such as Michael Dummett, Bob Hale, and Crispin Wright, who deny that *abstracta* exist in the same robust sense that indisputably existent objects exist, van Inwagen maintains that the predicate "exist" is univocal and that therefore abstract

> objects are just as real as anything else that truly exists. Abstract objects are for van Inwagen as real as elementary particles, people, and God Himself. But he thinks that they, like God, are *agenēta* (p. 198).

Dummett, Hale, and Wright are not lightweight Platonists. They are not lightweight Platonists because they are not Platonists of any kind. They are not Platonists of any kind because every kind of Platonism is a philosophical position and they have no position. They have no position because the words they use when they claim to be stating their position are meaningless.

My platonism is as lightweight as a platonism gets, so you might as well call it lightweight platonism. There's certainly no *other* use for the words 'lightweight platonism'.

Craig, of course, will dispute this. He calls my platonism "heavyweight" (he says) because it implies that "abstract objects are just as real as anything else that truly exists." But what does *that* mean? In particular, what does the word 'real' mean in that sentence? As far as I can see, there is only one possible answer to that question: 'real' means 'existent'. And, if that is so, 'just as real as anything else that truly exists' means 'just as existent as anything else that's truly existent.' That is to say, 'just as existent as anything else that's existent', owing to the fact that something is truly existent if and only if it's existent. And since 'existent' is an absolute adjective, the phrase 'just as existent as anything else that's existent makes no more sense than 'just as unique as anything else that's unique' or 'just as infinite as anything else that's infinite'. My platonism, therefore, does not imply that abstract objects are just as real as anything else that truly exists. And nor does it fail to imply that thesis—for there's no such thesis for it either to imply or fail to imply. The words 'are just as real as anything else that truly exists' are entirely meaningless.

Do I say that numbers, like God, are *agenēta*? Well, 'God is *agenētos*' is both true and one of the most important truths there is. 'The number six is *agenēton*' is both true and a very silly thing to say. In this respect, it is like 'The number six is impeccable' (in the theological sense of 'impeccable': incapable of sin). I suppose 'The number six is impeccable' has to count as true; I mean, there are lots of things capable of sin; make a complete list of them, and you'll find that the number six doesn't occur in your list. But being true doesn't save 'The number six is impeccable' from being a very silly proposition. In reading the passage I have quoted from Craig, I experience the same dizzying bewilderment I should feel if the following accusation were brought against me: "Van Inwagen contends that abstract objects have no causal powers. But the power to act is a causal power, and he is therefore committed to the position that abstract objects are incapable of action. But to sin is to act. Therefore, his doctrine implies that abstract objects are incapable of sin—that is, they are impeccable. *But only God is impeccable*!"

Summary

Craig has said that van Inwagen's platonism cannot be described as lightweight, owing to the fact that it implies that "abstract objects are just as real as anything else that truly exists." An argument is presented for the conclusion that 'just as real as anything else that truly exists' is meaningless. 'The number 6 is uncreated' is compared with 'The number 6 is incapable of sin'.)

6. Alternatives to Platonism

> Fortunately, there's just no necessity of making van Inwagen's theological compromise. Even without abandoning his favored criterion of ontological commitment, van Inwagen has plenty of options open to him other than Platonism (p. 198).

I, of course, insist that the "compromise" I am alleged to have made is a creature of Craig's imagination. I have no more made a theological compromise than Aquinas made a theological compromise when he wrote:

> But those things that do imply a contradiction do not fall within the scope of divine omnipotence, since they cannot have the nature of possible things.

As for the "options" Craig lays out—well, he may regard them as options, but I don't. I regard each of them as either meaningless, obviously wrong, or unable to account for the confidence we place in mathematics as a tool for describing and understanding the physical world. I do not have the space to treat each of them at the length it deserves, so I'll treat three of them much too briefly.

Absolute creationism: See "God and Other Uncreated Things" (Van Inwagen, 2009), particularly pp. 11–18.

Fictionalism: See the discussion of the position of the nominalist Norma (whose position is fictionalism, more or less) in Section 4 of my Reply in this volume—or my "Fictionalist Nominalism and Applied Mathematics." (Van Inwagen, 2014.)

Modal structuralism: A non-starter, because *all* **'were'/ 'would'-conditionals** with necessarily false antecedents are true.

> ### Conditionals
>
> **Conditionals** are "if-then" statements. There are at least two kinds of conditionals. For consider these two statements:
>
> If Booth didn't shoot Lincoln, someone else did.
> If Booth hadn't shot Lincoln, someone else would have.
>
> The first is obviously true—since *someone* shot Lincoln. The second is not obviously true, and may well be false. It is the second kind of conditional that interests us. The American philosopher David Lewis devised the symbol '□→' for the formal representation of those conditionals. Using this symbol, we represent 'If Booth hadn't shot Lincoln, someone else would have' as 'Booth didn't shoot Lincoln □→ someone else did', which we pronounce 'If it were the case that Booth didn't shoot Lincoln, it would be the case that someone else did'. We call 'Booth didn't shoot Lincoln' the **antecedent** of, and 'Someone else did' the **consequent** of, 'Booth didn't shoot Lincoln □→ someone else did'. A conditional of this kind is essentially the statement that if its antecedent *were* true then its consequent *would be* true. Hence, I call them **'were'/'would'-conditionals.** (They are also called "counterfactual" conditionals and "subjunctive" conditionals. Both those designations are flawed.)

There are *awfully* good arguments for that thesis. If you reject it, I have to wonder whether you're aware of those arguments. I can't help suspecting that you reject it simply because you've seen a few alleged counterexamples to it—a few sentences like 'If 10 were a prime number, it would be divisible by 3' and 'If I were a dolphin, I'd have legs'.

I'll write out just one of those "awfully good arguments." The core of the argument is a deduction of 'Bertrand Russell is the pope' from the premise '2 + 2 = 5' that (at least according to legend) was devised by Bertrand Russell. Legend ascribes various statements of the deduction to Russell, most of them something like this: "Assume that 2 + 2 = 5. Then, since 2 + 2 = 4, 4 = 5, Subtract 3 from each side of the equation, and you have 1 = 2, which is the same as 2 = 1. The pope and I are two—but, as we've seen, two is one. So the pope and I are one. That is, I am the pope."

And the argument is:

3. If there are any false 'were'/'would'-conditionals with necessarily false antecedents, one of them is

   2 + 2 = 5 □→ Bertrand Russell = the pope.[12] *Premise*

4. If the consequent of a 'were'/'would'-conditional is logically and mathematically deducible from its antecedent, that conditional is true. *Premise*

5. 2 + 2 = 5 *Assumption for the sake of argument*

6. 2 + 2 = 4 *Premise (from arithmetic)*

7. 4 = 5 *(5), (6) Substitution of identicals*

8. 1 = 2 *Subtract 3 from both sides of* (7)

9. 0 < the number of members of the set {Bertrand Russell, the pope} ≤ 2 *Premise*

10. 0 < the number of members of the set {Bertrand Russell, the pope} ≤ 1 *(8), (9) Substitution of identicals*

11. $\forall x \forall y$ (if 0 < the number of members of the set $\{x, y\} \leq 1$, then $x = y$) *Premise (from set theory)*

12. A technical note. I will assume a "Fregean" treatment of the semantics of terms: there is some object that is the "default" referent of every term that would otherwise lack a referent. This will secure the '0 <' in premise (9).

12. Bertrand Russell = the pope *(10), (11) Universal Instantiation, Modus ponens*

13. The consequent of '$2 + 2 = 5$ □→ Bertrand Russell = the pope' is logically and mathematically deducible from its antecedent *(5)–(12)*

14. '$2 + 2 = 5$ □→ Bertrand Russell = the pope' is a were/would conditional *Premise*

15. '$2 + 2 = 5$ □→ Bertrand Russell = the pope' is true. *(13), (14), (3)*

16. There are no false 'were'/'would'-conditionals with necessarily false antecedents. *(3), (15)*

Opponents of the conclusion of this argument, in response to similar arguments, tend to question premises like premise (4) of this argument. I would point out that (4) follows from these two principles:

> If *q* is logically and mathematically deducible from *p*, then if it were the case that *p* it would also have to be the case that *q*
>
> > If it were the case that *p* it would be the case that *q*
>
> follows from
>
> > If it were the case that *p* it would also have to be the case that *q*.

Summary

Craig has said that it's unnecessary to make "van Inwagen's theological compromise," owing to the fact that "van Inwagen has plenty of options open to him other than Platonism." It is denied that van Inwagen has made any compromise, and it is maintained that Craig's supposed alternatives are "meaningless, obviously wrong, or unable to account for the confidence we place in mathematics as a tool for describing and understanding the physical world." One of these alternatives, Modal Structuralism, is shown to have a false presupposition.

Chapter 6

Response to Peter van Inwagen's Reply

William Lane Craig

Contents

1. Introduction

When one agrees to enter into a formal debate, as we have, one assumes certain responsibilities that one does not otherwise have. These include presenting a prima facie case for one's answer to the question under debate and responding to the objections to one's case raised by one's interlocutor. Has Prof. van Inwagen discharged those responsibilities? Let's review.

Van Inwagen is engaged in an articulation and defense of his own version of the Indispensability Argument for Platonism. While in his Opening Statement he did a creditable job of presenting a case for his version of Premiss I of that argument, that is, for his criterion of ontological commitment, he failed to present a prima facie case for the existence of abstract objects, since he neglected to defend any version of Premiss II. His present Reply can now be read as remedying that lack, for he argues that there are literally true, simple sentences indispensably involving existential quantification over objects that could only be abstract objects. What we have, then, in van Inwagen's Reply is not a new argument for Platonism but rather a continuation of his Indispensability Argument

DOI: 10.4324/9781003008712-9

for Platonism, this time using numbers rather than properties as his paradigmatic abstract objects.[1]

Since interchanges such as ours can be difficult to track, I suggest to our readers that their understanding of our exchange will be enhanced if they consider our contributions in the following re-shuffled order:

1. Craig's Opening Statement: exposition of the Indispensability Argument, taxonomy of responses to the argument, prospects for anti-Platonist alternatives
2. Van Inwagen's Opening Statement and his Reply to Craig's Opening Statement: an extended case for Platonism, including a defense of favored versions of both premisses of the Indispensability Argument
3. Craig's Reply to van Inwagen's Opening Statement and his Response to van Inwagen's Reply: defense of anti-Platonist alternatives to both Premiss I and Premiss II of the Indispensability Argument
4. Van Inwagen's Response to Craig's Reply: **TBD**

Notwithstanding van Inwagen's extended case for Platonism, he has not yet shouldered his responsibility of responding to the objections offered by realist and anti-realist alternatives to Premiss II. Instead, he retorts, "I've spent a very long career refusing to let people tell me what I 'must' do in philosophy and I don't propose to begin taking such direction now." Well! Be that as it may, in a debate context it is incumbent upon a party to that debate to answer objections to his case, or he has not discharged his responsibilities in that debate. In his personal life van Inwagen is under no obligation to respond to anyone, but in entering into a debate situation he has assumed certain responsibilities. The "must" spoken of should be understood to arise from such a context.

In particular, we want to know why one should be a Platonist rather than, for example, a non-Platonic realist, such as a conceptualist. Or why not rather be an arealist? Or why not adopt some *ultima facie* strategy like modal structuralism or constructibilism, which paraphrase away alleged commitments to mathematical

1. Though, as he notes in his opening statement, van Inwagen reductively identifies numbers with properties.

objects, or figuralism, which takes mathematical discourse to be figurative rather than literal? Or how about some form of pretense theory like postulationalism, a popular anti-realist alternative among mathematicians? Or just good, ol' fictionalism rather than Platonism? Before we are persuaded to become Platonists, we want to know why we should not embrace any of these alternatives. Thus, wholly apart from the objections that I raised in my Reply to van Inwagen's defense of Premiss I, which he cannot be expected to have answered yet, he really needed to respond to the proffered alternatives to his Premiss II.

Now van Inwagen speaks disparagingly of Premiss II: "I'd rather take Craig's 'Premiss II' apart and lay its pieces on the table and explain to anyone who wants to listen why I'd never use a premise put together from such mutually ill-adjusted parts." Fine, I'm listening! What parts of Premiss II does he reject and what would he replace them with in his preferred version of this step in his Indispensability Argument? Van Inwagen hints tantalizingly, "my presentation of my arguments for 'lightweight platonism' in the opening statement pretty clearly display how I'd reply to any of the 'defeaters' Craig mentions—generally by saying either, 'That doesn't apply to my position' or 'That's meaningless.'"

I am baffled. Recall that for van Inwagen lightweight Platonism is just the view that abstract objects are one and all causally effete. Of his introduction to lightweight Platonism in Section 2 of his Opening Statement, he says, "Its primary purpose is not to defend lightweight platonism but rather to convince the reader that the concept 'property' that figures in lightweight platonism is a concept that the Fathers of the Church did not have." Fine; so where in his Opening Statement do we come to the argument for lightweight Platonism? **Sections 3–8** of his essay "are devoted," he says, to my charge that van Inwagen's "reasons for believing that there are abstract objects essentially incorporate a wrong 'meta-ontology'." Right; then follows his defense of his criterion of ontological commitment, or Premiss I of his Indispensability Argument. Now that's the whole Opening Statement, and I haven't found therein any defense of the view that abstract objects exist.

Saying of alternative views either "That doesn't apply to my position" or "That's meaningless" is a nice slogan but is far, far too quick to turn back the force of these various realist and anti-realist alternatives to Platonism, which have been articulated and defended by their proponents in full consciousness of van Inwagen's neo-Quinean position. Is Plantinga's conceptualism meaningless? Inapplicable to van Inwagen's position? How about Hellman or Chihara's **paraphrastic strategies**? Since they adopt the Quinean strategy for shedding ontological commitments, they must apply to van Inwagen's position; so are they meaningless? Why? Or Field or Balaguer's fictionalism—since they accept a neo-Quinean criterion of ontological commitment, how could their position be inapplicable to van Inwagen's position? But neither is fictionalism meaningless by van Inwagen's lights, since he elsewhere argues against it. That's what we want here: reasons why these various alternatives should be rejected in favor of Platonism.

Now in one sense that's what van Inwagen's present Reply offers us. It extends his Indispensability Argument by claiming that there are literally true sentences indispensably involving quantification over or reference to abstract objects; otherwise the applicability of mathematics becomes inexplicable. So these other anti-Platonic views are implicitly ruled out. The problem is, the theorists who defend these various views are fully aware of this argument and have offered responses to it. One cannot responsibly just ignore those responses.

2. Anti-Platonist Reponses to Premiss I

This is not even to mention views which reject Premiss I of the Indispensability Argument, which I have re-defended in my Reply to van Inwagen's Opening Statement. Since I favor neutralism, I look forward to reading his response to neutralism in his final Response. Piggy-backing on the work of free logicians, neo-Meinongians, and neutralists, I've articulated a view of quantification as ontologically neutral and of reference as an intentional activity of agents which allows us to affirm the mathematical truths which van Inwagen champions in his present Reply. The burden of his Reply is to show that the applicability of mathematics to the world is inexplicable apart from mathematical truth. Such a claim is, however, neutral with respect to the debate between Platonism and anti-Platonism, for conceptualists, free logicians, neo-Meinongians, *ultima facie* strategists, and neutralists all affirm mathematical truth. His argument has purchase only against views that reject mathematical truth.

For that reason van Inwagen greatly misleads his reader when he states that in his Reply he will argue for the thesis "that if numbers and other mathematical objects do not exist, then the applicability of mathematics to the world about us (which cannot be disputed) is a mystery."[2] For what van Inwagen actually argues for is the thesis that if numerical and other mathematical sentences are not true, then the applicability of mathematics is a mystery. His is an argument for mathematical truth, not mathematical objects. It is only in conjunction with his criterion of ontological commitment that mathematical truth implies the reality of mathematical objects. As a neutralist who believes in mathematical truth, I'm unfazed by his response.

Moreover, it needs to be kept firmly in mind that van Inwagen and the neutralist agree that sentences of ordinary English involving quantificational expressions or singular terms are susceptible of both a lightweight and a heavyweight reading and, moreover, that a sentence may be true when taken in a lightweight sense but false when taken in a heavyweight sense. Thus, for example, the sentence "The crack in the Liberty Bell is 62.23 cm in length" is true when taken in a lightweight sense but false when taken in a heavyweight sense. For the expression "the crack in the Liberty Bell" does not, by van Inwagen's lights, denote an existent object.[3] Now the neutralist says the same thing about the sentences of applied mathematics. Jody Azzouni, for example, is adamantly opposed to fictionalism with respect to such sentences, insisting like van Inwagen that the applicability of mathematics becomes mysterious if we deny the truth of such sentences.[4] So Azzouni would concur with the truth of all the premisses of the Expanded Farmer's Argument, including those inferred by UI and EG.[5] But Azzouni is talking about the truth of such sentences in a lightweight sense. Van Inwagen's argu-

2. Again, "I will suggest that anyone who denies the existence of mathematical objects . . . must regard the applicability of mathematics to the world as a mysterious, or at any rate an inexplicable, brute fact."
3. The expression is therefore, by van Inwagen's lights, not really a singular term (a name).
4. Jody Azzouni, *Deflating Existential Consequence: A Case for Nominalism* (Oxford: Oxford University Press, 2004), Chapters 1–2; *cf.* idem, "Evading Truth Commitments: The Problem Reanalyzed," *Logique & Analyse* 206 (2009): 139–76.
5. Van Inwagen's view of EG and UI is interestingly similar to the view of free logicians that the validity of EG and UI require first the affirmation that the relevant object exists. Just as we cannot know whether UI and EG are applicable to sentences involving the expression "the crack in the Liberty Bell" until we know that such an object exists, so we cannot know whether sentences involving numerical expressions are susceptible to UI or EG unless and until we know that numbers exist. But that

ment is, after all, expressed in ordinary English, not his Revised English. Translated into Revised English, such sentences are not, in Azzouni's view, true.

What van Inwagen needs to show, therefore, is that the truth of his sentences in Revised English is necessary for mathematics' applicability. There need to be abstract objects answering to mathematical terms in order for mathematics to be useful in calculating the stock density of a field. But now the anti-Platonist's objection seems to go through with a vengeance. As Mark Balaguer complains, the idea that Platonism somehow accounts for the applicability of mathematics "is actually very counterintuitive. The idea here is that in order to believe that the physical world has the nature that empirical science assigns to it, I have to believe that there are causally inert mathematical objects, existing outside of spacetime," an idea which is inherently implausible.[6] For the Platonist, the fact that physical reality behaves in line with the dictates of acausal mathematical entities existing beyond space and time is, in the words of Mary Leng, "a happy coincidence."[7] Truth in the lightweight sense is sufficient for the inferences that van Inwagen extols; truth in the heavyweight sense involving causally unconnected abstract objects adds nothing.

3. Anti-Platonist Reponses to Premiss II

But now, to return to anti-Platonist alternatives to Premiss II, we want to ask whether van Inwagen has demonstrated that the reality of abstract mathematical objects is necessary for applied mathematical reasoning. Given van Inwagen's lack of engagement with alternative views, that seems dubious.

3.1 Views Not Affirming Mathematical Truth

Consider, first, views that do not regard mathematical sentences as true. Some fictionalists, for example, Hartry Field, take the radical line that the truth of (heavyweight) mathematical sentences is not required for our understanding of the world because mathematics is ultimately dispensable. On Field's view, a nominalization of physical

is precisely the question under dispute. Accordingly, the Expanded Farmer's Argument cannot provide a non- question-begging reason for being a Platonist.

6. Balaguer, Platonism and Anti-Platonism in Mathematics, p. 136.
7. Leng, Mathematics and Reality, p. 239.

science is vital to a successful anti-realism, for the affirmation of fictionalism *tout court* leaves one without an account of how natural science can, if it is riddled with falsehoods, accurately picture the world.[8] Most fictionalists demur, however, holding that the nominalization of science is not essential to the defeat of the Quinean objection. Balaguer, for example, is willing to concede Quinean claims about the indispensability of mathematics to physical science, while maintaining realism about empirical science and fictionalism about mathematics. Balaguer defends what he calls nominalistic scientific realism, the view that the nominalistic content of empirical science is (mostly) true, while its Platonistic content is fictional.[9] The Platonistic content is used non-assertorically as a useful instrument to get at the nominalistic content of empirical science.

But what about van Inwagen's complaint that, given the indispensability of mathematical terms, arguments for even nominalistically acceptable conclusions are undermined, leaving us with no reason to think that truth should be ascribed to their conclusions? Balaguer admits that he has no explanation why, on fictionalism, mathematics is applicable to the physical world or why it is indispensable in empirical science. He just observes that neither can the Platonist answer such "why" questions. That answer must be sought elsewhere.

I think Balaguer is right. Van Inwagen's "explanation" of mathematics' applicability is actually very superficial. It consists merely in the soundness of reasoning with premisses involving applied mathematics. But, as I say, van Inwagen's Expanded Farmer's Argument is not formulated in Revised English. It is striking, for example, that in his list of quantificational phrases on p. 40 we do not find the characteristic existential quantificational phrase of Revised English "something is such that."[10] Van Inwagen's premisses are evidently formulated

8. Field, *Science without Numbers,* pp. 1–2. This is precisely the objection pressed by van Inwagen, who assumes that Field's nominalization project will not go through.
9. Mark Balaguer, *Platonism and Anti-Platonism in Mathematics* (New York, NY: Oxford University Press, 1998), pp. 130*ff.*
10. As van Inwagen says, "Outside the Ontology Room—in the ordinary business of life—no such rule is in force. (I don't suppose that any speaker engaged in the ordinary business of life would utter a sentence that was the result of appending a singular count-noun to 'Something is such that it is a'. If I did hear a sentence of that form uttered in a florist's shop or a law office or a lecture on the history of Chinese pottery, I'm not sure what I should conclude." That goes for sentences uttered by sheep farmers as well.

in ordinary English, so that we needn't stumble at pseudo-entities like lengths-in-meters and stock-densities-in-sheep-per-acre. If we accordingly give his premisses a lightweight reading, not only the neutralist, but even the hard-nosed fictionalist may affirm their truth. For the fictionalist regards as false only sentences in which existential quantifiers and singular terms are devices of ontological commitment, i.e., sentences of Revised English. So paradoxically even the fictionalist can affirm the truth of van Inwagen's premisses, so long as one doesn't translate them into Revised English, and thus "explain" the applicability of mathematics. Van Inwagen needs to show that the existence of mind-independent, abstract objects existing in the real world somehow contributes to the explanation of mathematics' applicability. Fictionalists are understandably sceptical of such a claim.

All this still leaves us wondering why such mathematical reasoning is effective in the physical world. Neither the Platonist nor the fictionalist as such has a deep explanation to offer for the applicability of mathematics. It seems to me that the explanation of mathematics' applicability is to be found, not in some property of mathematics, but in something about the nature of the physical world. The concrete world is constructed in such a way that standard mathematics, in contrast to non-standard theories, is useful and reliable. Of course, this answer leaves unanswered the further question as to why the physical world is such as to make standard mathematical sentences applicable to the world—a question to which I shall return momentarily.

3.2 Views Affirming Mathematical Truth

Most of the anti-realist alternatives offered in response to Premiss II in fact affirm the truth of mathematical sentences. Think, for example, of various *ultima facie* strategies for dealing with the Platonistic commitments of such sentences. Rather than take on Field's project of reformulating science so as to rid it of mathematical sentences, these theorists offer a reformulation of mathematical sentences so that they no longer involve quantification over or reference to mathematical objects. In **Section 2.2** of my Reply, I raised a number of problems with van Inwagen's attempt to exclude acceptable paraphrases of such sentences, problems that merit some response. Hellman and Chihara have shown us how to paraphrase the Expanded Farmer's Argument in such a way as to retain the truth of its premisses without quantifying over mathematical objects. That's all that van Inwagen's argument for applicability requires. Yablo has shown how such sentences can be construed figuratively, so that their truth does not imply the existence of mathematical objects. What's the matter with that alternative?

Then there are all the other realist and anti-realist alternatives to Platonism that affirm the truth of mathematical sentences and are therefore immune to van Inwagen's argument for Platonism from the truth of mathematical sentences. Until van Inwagen replies to these alternatives defended by various prominent philosophers of mathematics, his argument suffers defeat.

4. Explaining the Applicability of Mathematics

So "What explains the fact that applied mathematics is possible? What is the ground of the trust that we place in the answers to our questions about the world that calculation and mathematical reasoning provide?" More specifically, "Why was the Farmer's reasoning trustworthy?" According to van Inwagen, the answer is

> The Farmer's Argument is an enthymematic abbreviation of the Farmer's Argument (Expanded), which is a sound argument. That the figure the farmer's calculation produced was correct has, and needs, no more explanation than this: The conclusion of a sound argument must be true.

I have said that this explanation is superficial. For it does not explain why physical phenomena are such that mathematical reasoning applies to them. This question becomes acute as one rises from elementary arithmetic truths such as the farmer uses to the daunting equations of theoretical physics. Here we confront what Eugene Wigner famously called "the unreasonable effectiveness of mathematics."[11]

11. Eugene Wigner, "The Unreasonable Effectiveness of Mathematics in the Natural Sciences," in *Communications in Pure and Applied Mathematics* 13/1 (New York: John Wiley & Sons, 1960).

4.1 Wigner's Quandry

As a mathematical physicist, Wigner was surprised at the effectiveness of mathematics in describing physical phenomena. He observes that mathematical concepts turn up in "entirely unexpected connections" in physics and often permit "an unexpectedly close and accurate description" of the phenomena in these connections.[12] Wigner shows by numerous examples that mathematics plays a central role in the formulation of successful laws of physics, a conclusion which no one, I think, would contest.

Wigner's key question comes in the section of his paper entitled, "Is the Success of Physical Theories Truly Surprising?" Here Wigner stresses the a priori nature of mathematical inquiry, especially of the mathematics that is so valuable in physics:

> whereas it is unquestionably true that the concepts of elementary mathematics and particularly elementary geometry were formulated to describe entities which are directly suggested by the actual world, the same does not seem to be true of the more advanced concepts, in particular the concepts which play such an important role in physics. . . . Most more advanced mathematical concepts, such as complex numbers, algebras, linear operators, Borel sets - and this list could be continued almost indefinitely - were so devised that they are apt subjects on which the mathematician can demonstrate his ingenuity and sense of formal beauty.[13]

The "principal point" which will be relevant to the uncanny effectiveness of mathematics is that mathematicians are not bound by customary axiomatic concepts but are free to define new concepts with a view, not of applicability or scientific utility, but of "permitting ingenious logical operations which appeal to our aesthetic sense both as operations and also in their results of great generality and simplicity."[14] That historically there has been a cross-pollination between physics and mathematics,

12. Ibid.
13. Ibid., pp. 2–3.
14. Ibid.

physics sometimes spurring developments in mathematics,[15] does not nullify Wigner's point. Wigner finds a particularly striking example in complex numbers.

> Certainly, nothing in our experience suggests the introduction of these quantities. Indeed, if a mathematician is asked to justify his interest in complex numbers, he will point, with some indignation, to the many beautiful theorems in the theory of equations, of power series, and of analytic functions in general, which owe their origin to the introduction of complex numbers. The mathematician is not willing to give up his interest in these most beautiful accomplishments of his genius.[16]

Moreover, I should add, when we reflect that mathematical objects, even if they exist, are causally effete, it is surprising that such objects should be significantly effective in physics. Even those who maintain that mathematics plays an explanatory role in science recognize that such cases involve acausal explanation.[17] The "explanation" amounts to nothing more than the broadly logical necessity of mathematical truths, e.g., Mother cannot divide 23 strawberries among her three children evenly because 23 is not divisible by 3 without remainder. Van Inwagen's illustration of calculating the stock density of a field is precisely an example of this sort. Such acausal explanations do nothing to account for the fact that mathematics is so effective in describing physical phenomena. Indeed, the abstractness of mathematical objects underlines in general terms,

---

15. Ivor Grattan-Guiness, "Solving Wigner's Mystery: The Reasonable (Though Perhaps Limited) Effectiveness of Mathematics in the Natural Sciences," *Mathematical Intelligencer* 30/3 (2008): 7–17.
16. Wigner, "Unreasonable Effectiveness of Mathematics," p. 3.
17. For example, Manfred R. Schroeder, "The Unreasonable Effectiveness of Number Theory in Physics, Communication and Music," *Proceedings of Symposia in Applied Mathematics* 46 (1992): 1–19, http://dx.dol.org/10.1090/psapm/046/1195839; Alan Baker, "Are there Genuine Mathematical Explanations of Physical Phenomena?" *Mind* 114/454 (2005): 223–38; Marc Lange, "What Makes a Scientific Explanation Distinctively Mathematical?," *British Journal for the Philosophy of Science* 64/3 (2013): 485–511. Lange writes, "these explanations explain not by describing the world's causal structure, but roughly by revealing that the explanandum is more necessary than ordinary causal laws are. . . . These necessities are stronger than causal necessity, setting distinctively mathematical explanations apart from ordinary scientific explanations" (Lange, "What Makes a Scientific Explanation Distinctively Mathematical?," p. 491).

not just in isolated examples, how surprising their applicability to physical phenomena is.

Accordingly, we may formulate Wigner's argument as follows:

1. Mathematical concepts arise from the aesthetic impulse in humans and have no causal connection to the physical world.
2. It would be surprising to find that what arises from the aesthetic impulse in humans and has no causal connection to the physical world should be significantly effective in physics.
3. Therefore, it would be surprising to find that mathematical concepts should be significantly effective in physics.
4. The laws of nature can be formulated as mathematical descriptions (concepts) which are often significantly effective in physics.
5. Therefore, it is surprising that the laws of nature can be formulated as mathematical descriptions that are often significantly effective in physics.

Given that something surprising merits prima facie an explanation, we wonder as to the explanation of the fact that the laws of nature can be formulated as mathematical descriptions that are often significantly effective in physics.

Wigner, despite his repeated characterization of the applicability of mathematics to the physical world as "a miracle," in the end regarded it as a mystery. He concluded "that the enormous usefulness of mathematics in the natural sciences is something bordering on the mysterious and that there is no rational explanation for it."[18] Wigner would expand van Inwagen's claim to read, "If we say we trust these methods to give the right answer because and *only* because we know that they have faultless track records or because we know that the laws are true, our position implies that the trustworthiness of mathematics is a brute fact, a mystery."

4.2 Theism and the Applicability of Mathematics

Wigner, however, never actually considered in his essay whether the applicability of mathematics might not be a literal miracle. He considered at most naturalistic explanations of it and, finding

18. Wigner, "Unreasonable Effectiveness of Mathematics," p. 2.

none to be satisfactory, therefore concluded, "The miracle of the appropriateness of the language of mathematics for the formulation of the laws of physics is a wonderful gift which we neither understand nor deserve."[19] But suppose we take the theistic hypothesis seriously. Since the question of mathematics' applicability to the physical world is already a metaphysical, not a physical, question, there can be no objection stemming from the corner of methodological naturalism to considering a metaphysical answer to a metaphysical question.

Theists will have a considerably easier time, I think, explaining the applicability of mathematics than will naturalists. Theists hold that there is a personal, transcendent being (a.k.a. God) who is the Creator and Designer of the universe. Naturalists hold that all that exists concretely is space-time and its physical contents. Now whether one is a realist or an anti-realist about mathematical objects, it appears that the theist enjoys a considerable advantage over the naturalist in explaining the uncanny effectiveness of mathematics.

4.2.1 Realism About Mathematical Objects and Theism

Consider first realism's take on the applicability of mathematics to the world. For the *non-theistic* realist, the fact that physical reality behaves in accord with the dictates of acausal mathematical entities existing beyond space and time is inexplicable. For consider: If, *per impossibile*, all the abstract objects in the mathematical realm were to disappear overnight, there would be no effect on the physical world. This is simply to underscore the fact that abstract objects are causally inert. The idea that realism somehow accounts for the applicability of mathematics is thus inherently implausible.

By contrast, the *theistic* realist can argue that God has fashioned the world on the structure of the mathematical objects He has chosen. This is essentially the view that Plato defended in his dialogue *Timaeus*.[20] Plato draws a fundamental distinction between the realm of static being (that which ever is) and the realm of temporal becoming (that which is ever becoming). The realm of becoming

19. Ibid., p. 14.
20. *Timaeus* 3–4.

comprises primarily physical objects, while the static realm of being comprises logical and mathematical objects. God looks to the realm of mathematical objects and models the world on it. The world has its mathematical structure as a result.

The main objection confronting this view is theological: the realm of mathematical objects is thought to exist independently of God, so that God is not the sole ultimate reality. But, as we have seen, some Christian realists advocate absolute creationism, the view that the realm of abstract objects, including mathematical objects, though necessary in its existence, is nonetheless causally dependent upon God,[21] while others advocate divine conceptualism, the view that mathematical and other putative abstract objects are thoughts of various sorts in the mind of God and so dependent upon God for their being.[22] Thus, the realist who is a theist has a considerable advantage over the naturalistic realist in explaining why mathematics is so effective is describing the physical world.

4.2.2 Anti-realism About Mathematical Objects and Theism

Now consider anti-realism of a *non-theistic* sort. Leng says that on anti-realism relations which are said to obtain among pretended mathematical objects just mirror the relations obtaining among things in the world, so that there is no happy coincidence. Philosopher of physics Tim Maudlin muses, "The deep question of why a given mathematical object should be an effective tool for representing physical structure admits of at least one clear answer:

21. This is the option advocated by Thomas Morris and Christopher Menzel. See Thomas V. Morris and Christopher Menzel, "Absolute Creation," *American Philosophical Quarterly* 23/4 (1986): 353–62; Christopher Menzel, "Theism, Platonism, and the Metaphysics of Mathematics," *Faith and Philosophy* 4/4 (1987): 365–82; Christopher Menzel, "God and Mathematical Objects," in *Mathematics in a Postmodern Age*, ed. Russell W. Howell and W. James Bradley (Grand Rapids, MI: Eerdmans, 2001), pp. 65–97.
22. This is the option advocated by Alvin Plantinga and defended most extensively by Greg Welty. See Alvin Plantinga, "Theism and Mathematics," *Theology and Science* 9/1 (2011): 27–33; Alvin Plantinga, *Where the Conflict Really Lies: Science, Religion, and Naturalism* (Oxford: Oxford University Press, 2011), pp. 284–86; Greg Welty, "Theistic Conceptual Realism," in *Beyond the Control of God?: Six Views on the Problem of God and Abstract Objects*, edited by Paul M. Gould (London: Bloomsbury, 2014), pp. 81–96.

because the physical world literally has the mathematical structure; the physical world is, in a certain sense, a mathematical object."[23] Well and good, but what remains wanting on naturalistic anti-realism is an explanation *why* the physical world should exhibit so elegant and stunning a mathematical structure in the first place. After all, there is no necessity that a physical world exist at all, in which case mathematical truths would not have been descriptive of the physical world. Perhaps the universe, in order to exist, had to have *some* mathematical structure—though couldn't the world have been a structureless chaos?[24]—but that structure might have been describable by elementary arithmetic. The Expanded Farmer's Argument depends on that sort of structure. But, as Wigner is at pains to emphasize, modern physics shows the physical world to be breathtakingly mathematically elegant. By using as his examples laws of nature which are fearsomely advanced mathematically, Wigner already forced the question to a higher plane.

By contrast, the *theistic* anti-realist has a ready explanation of the applicability of mathematics to the physical world: God has created the universe according to a certain model which He had in mind. There are any number of models He might have chosen. The laws of nature have the mathematical form they do because God has chosen to create the world according to the abstract model He had in mind. This was the view of the first century Jewish philosopher Philo of Alexandria, who maintained in his treatise *On*

23. "On the Foundations of Physics," July 5, 2013, 3:16, https://www.3-16am.co.uk/articles/on-the-foundations-of-physics
24. Albert Einstein thought so:

> One should expect a chaotic world which cannot be grasped by the mind in any way. One could (yes *one should*) expect the world to be subjected to law only to the extent that we order it through our intelligence. . . . By contrast, the order created by Newton's theory of gravitation, for instance, is wholly different. Even if the axioms of the theory are proposed by man, the success of such a project presupposes a high degree of ordering of the objective world, and this could not be expected *a priori*. That is the 'miracle' which is being constantly reinforced as our knowledge expands.
>
> (Albert Einstein, letter to Maurice Solovine, March 30, 1952, in Albert Einstein, *Letters to Solovine*, with an Introduction by Maurice Solovine, trans. Wade Baskin [New York, NY: Philosophical Library, 1987]. pp. 132–33).

I am indebted to Melissa Cain Travis for this reference.

the Creation of the World that God created the physical world on the mental model in His mind.[25] Philo's view can be interpreted as either a conceptualist realism or as an anti-realism, for he says that the intelligible world may be thought of as either formed by the divine Logos or, more reductively, as the Logos itself as God is engaged in creating.

Thus, the theist—whether he be a realist or an anti-realist about mathematical objects—has the explanatory resources to account for the otherwise unreasonable effectiveness of mathematics in physical science—resources which the naturalist lacks.

4.3 Wigner's Argument Extended

We may thus extend Wigner's argument:

6. Therefore, the fact that the laws of nature can be formulated as mathematical descriptions that are often significantly effective in physics merits explanation.
7. Theism provides a better explanation of the fact that the laws of nature can be formulated as mathematical descriptions that are often significantly effective in physics than does atheism.
8. Therefore, the fact that the laws of nature can be formulated as mathematical descriptions that are often significantly effective in physics provides evidence for theism.

Theism provides a substantive explanation of the applicability of mathematics, an explanation which is independent of the debate between realists and anti-realists over the existence of putative abstract objects. Since van Inwagen is interested in explaining the applicability of mathematics, he should avail himself of the explanatory resources of the theism which both he and I share.

5. Summary and Conclusion

To wrap up, I think that as a result of Prof. van Inwagen and my very interesting exchange, we can safely draw the following conclusions:

25. On the Creation of the World 16–20; 24.

1. Ordinary English sentences involving existential quantification and singular terms often have both metaphysically lightweight and metaphysically heavyweight readings.
2. Existential quantification over and singular terms referring to certain things are therefore not automatically means of ontological commitment.
3. We may use a fragment of an artificial language like Revised English to make our ontological commitments clear.
4. Our inability to provide a paraphrase in Revised English of lightweight claims made in ordinary English is of no significance.
5. There are nonetheless ways of paraphrasing in Revised English ordinary English sentences quantifying over or referring to abstract mathematical objects like numbers so as not to involve ontological commitment to such objects.
6. There are legitimate ways of understanding mathematical discourse as figurative or make-believe and, hence, ontologically non-committing.
7. The truth of mathematical sentences does not explain the applicability of mathematics to physical phenomena.
8. Whether one is a realist or anti-realist about mathematical objects, theism provides the best explanation of the applicability of mathematics to physical phenomena.
9. We so far forth have no good reason to believe in the reality of mathematical objects like numbers.
10. Given the presumption against such bizarre objects as numbers the dispensability of mathematical objects provides grounds for thinking that such objects do not exist.

Finally, I want to close on a light-hearted note. Prof. van Inwagen is a great fan of the so-called double dactyl, a light-verse form invented by the American poets John Hollander and Anthony Hecht, and has even published such amusing poems of his own. As a playful token of my esteem I offer the following:[26]

"Peter van Inwagen"
Higgledy-piggledy,

26. My poem requires us to put the stress on the first syllable of his last name: INwagen, which is the way in which everybody (apart from Peter) pronounces his name.

Peter van Inwagen,
Meta-ontology
borrowed from Quine,

Finds that some sentences
Quantificational
Foist on us entities
Fain left behind.

Numbers and properties,
Shapes geometrical,
Fictional characters,
infinite sets--!

They are more real, sayeth
Peter van Inwagen,
E'en than the furniture
On which he sits!

Further Readings

Suggested Readings (William Lane Craig)

Craig, William Lane. *God over All: Divine Aseity and the Challenge of Platonism.* Oxford: Oxford University Press, 2016.

Frege, Gottlob. *The Foundations of Arithmetic: A Logico-Mathematical Enquiry into the Concept of Number,* translated by J. L. Austin. 2nd rev. ed. Evanston, IL: Northwestern University Press, 1968.

Gould, Paul M., ed. *Beyond the Control of God?: Six Views on the Problem of God and Abstract Objects.* Bloomsbury Studies in Philosophy of Religion. London: Bloomsbury, 2014.

Potter, Michael. *Set Theory and Its Philosophy: A Critical Introduction.* Oxford: Oxford University Press, 2004.

Indispensability Argument

Balaguer, Mark. "Platonism in Metaphysics." In *The Stanford Encyclopedia of Philosophy*. Summer 2009 ed. Article published June 21, 2009. http://plato.stanford.edu/archives/sum2009/entries/platonism/.

Van Inwagen, Peter. "God and Other Uncreated Things." In *Metaphysics and God: Essays in Honor of Eleonore Stump*, edited by Kevin Timpe, 3–20. London: Routledge, 2009.

Van Inwagen, Peter. "A Theory of Properties." In *Oxford Studies in Metaphysics*, edited by Dean Zimmerman, 1: 107–38. Oxford: Oxford University Press, 2004.

Absolute Creationism

Morris, Thomas V., and Christopher Menzel. "Absolute Creation." *American Philosophical Quarterly* 23, no. 4 (1986): 353–62.

Physicalism

Franklin, James. *An Aristotelian Realist Philosophy of Mathematics: Mathematics as the Science of Quantity and Structure.* Basingstoke, Hampshire: Palgrave Macmillan, 2014.

Conceptualism

Welty, Greg. "Theistic Conceptual Realism: The Case for Interpreting Abstract Objects as Divine Ideas." D.Phil. thesis, University of Oxford, 2006.

Arealism

Maddy, Penelope. *Defending the Axioms: On the Philosophical Foundations of Set Theory.* Oxford: Oxford University Press, 2011.

Neutralism

Azzouni, Jody. *Deflating Existential Consequence: A Case for Nominalism.* Oxford: Oxford University Press, 2004.

Azzouni, Jody. "Ontological Commitment in the Vernacular." *Noûs* 41, no. 2 (2007): 204–26.

Free Logic

Lambert, Karel. *Free Logic: Selected Essays.* Cambridge: Cambridge University Press, 2003.

Ultima Facie Strategies

Chihara, Charles S. *Constructibility and Mathematical Existence.* Oxford: Clarendon Press, 1990.

Hellman, Geoffrey. *Mathematics without Numbers: Towards Modal-Structural Interpretation.* Oxford: Oxford University Press, 1989.

Yablo, Stephen. "Go Figure: A Path Through Fictionalism." In *Figurative Language,* edited by Peter A. French and Howard K. Wettstein. Midwest Studies in Philosophy, 25: 72–102. Oxford: Blackwell, 2001.

Fictionalism

Balaguer, Mark. *Platonism and Anti-Platonism in Mathematics.* New York: Oxford University Press, 1998.

Neo-Meinongianism

Routley, Richard. *Exploring Meinong's Jungle and Beyond: An Investigation of Noneism and the Theory of Items.* Canberra: Research School of Social Sciences, Australian National University, 1979.

Pretense Theory

Leng, Mary. *Mathematics and Reality.* Oxford: Oxford University Press, 2010.

Walton, Kendall L. *Mimesis as Make-Believe: On the Foundations of the Representational Arts.* Cambridge, MA: Harvard University Press, 1990.

Suggested Readings (Peter van Inwagen)

The following works constitute the history of the authors' debates on the topics of the present book

"God and Other Uncreated Things" (Van Inwagen, 2009)

God over All: Divine Aseity and the Challenge of Platonism (Craig, 2016)

"God and Abstract Objects" (Craig, 2015a)

"Response to Bridges and Van Inwagen" (Craig, 2015b)

"Did God Create Shapes?" (Van Inwagen, 2015a)

"A Reply to Dr Craig" (Van Inwagen, 2015b)

"Response to Van Inwagen and Welty" (Craig, 2019)

"Response to William Lane Craig's *God Over All*" (Van Inwagen, 2019)

Two books that defend positions essentially the same as Craig's are *Ontology and the Ambitions of Metaphysics* (Hofweber, 2016) and *Deflating Existential Consequence: A Case for Nominalism* (Azzouni, 2004).

Robert M. Adams's *What Is, and What Is in Itself* (Oxford: Oxford University Press, 2021) contains an important defense of the position that abstract objects exist only as objects of the thoughts of concrete rational beings.

More detailed and technical discussions of the positions I defend in the present volume can be found in "A Theory of Properties" (Van Inwagen, 2004) and "Inside and Outside the Ontology Room" (Van Inwagen, 2014b). My reasons for rejecting "fictionalism" are set out in detail in "Fictionalist Nominalism and Applied Mathematics" (Van Inwagen, 2014a).

The ultimate source of my convictions about the nature of being and the existence of abstract objects is one of the most admired and frequently cited philosophical essays of the twentieth century, W. V. O. Quine's "On What There Is." This classic essay was first published in *The Review of Metaphysics* in 1948 (Vol. 2, pp. 21–38), but is probably more easily found in the collection *From a Logical Point of View* (Cambridge, MA: Harvard University Press, 1953).

Two important essays belonging to the philosophical tradition founded by "On What There Is" are "Ontological Commitment" (Church, 1958) and "Holes" (Lewis and Lewis, 1983). They are difficult, and the beginning student will almost certainly need help and guidance to make it all the way through either of them, but those who are genuinely interested in the topics debated in this book will find that the effort put into understanding them was well worth it.

Alvin Plantinga's *Does God Have a Nature?* (Milwaukee, WI: Marquette University Press, 1980) is an important contribution to our understanding of the relationship between God and abstract objects. It is written from a point of view close to mine.

Glossary

A

***a se*:** Latin expression for self-existence; *lit.*, by itself
absolute creationism: the view that God has created all abstract objects
abstract objects: immaterial entities which are essentially causally powerless
anti-Platonism: the view that abstract objects do not exist
anti-realism: the view with respect to a certain thing that the thing does not exist
applicability of mathematics: mathematics' utility in describing the physical world
arealism: the view that there is no fact of the matter concerning something's existence
aseity: self-existence

B

bootstrapping objection: an objection alleging a vicious circularity

C

conceptualism: the view that supposedly abstract objects are really concrete thoughts of God
concrete objects: objects which have causal powers, such as minds and material objects
conservation: God's preserving things in being moment by moment

contingent: possible but not necessary
conventionalism: the view that answers to metaphysical questions are wholly arbitrary, having no truth value
creatio ex nihilo: Latin expression for creation without a material cause (*lit.*, creation out of nothing)
criterion of ontological commitment: a criterion revealing to us just what existential commitments our assertions have; it tells us what we have to believe exists in order to hold to the truth of our assertion

D

deflationary: minimalist; reducing one's ontological commitments; opposite of inflationary
denote: a term's singling out some object which has the properties predicated by the statement in which the term occurs; the object is the denotation of the term

E

entity: a thing; a being; an object
essential: belonging to a thing's very nature
eternal: having neither beginning nor end in time; permanent
exemplification: a relation between a thing and a property which the thing has; *e.g.*, Fido exemplifies the property of *brownness*
exemplify: to have (a property)
Existential Generalization: a logical rule permitting us to infer from "*a* is *F*" that "There is something that is *F*" (or, "Something is *F*")
existentially loaded: carrying ontological commitment
existentially quantified statement: an assertion about some of the members of a class, usually indicated by words like "some," "there is/are," "at least one," and so forth
existential quantifier: a logical operator which makes a statement an existentially quantified statement. Ordinary language expressions like "there is/are," "some," "at least one," and so forth, are informal quantifiers; in formal or symbolic logic the operator "$\exists$" is a formal quantifier symbolizing such informal expressions.
existential sentence: an existentially quantified sentence

extensional context: sentence phrases which have two characteristics: (i) singular terms referring to the same entity can be switched without affecting the sentence's truth value; (ii) one can quantify into such contexts from the outside

external questions: questions posed in a language not used in an adopted linguistic framework

F

fictionalism: the view that abstract object discourse is false or untrue

fictionally true: prescribed to be imagined as true

figuralism: the view that abstract object discourse is figurative language

figurative language: non-literal language, *e.g.*, metaphorical language

Form: for Plato transcendent, abstract objects on which God patterns the sensible world; for Aristotle immaterial entities which imbue matter to make it into some specific thing such as a rock or a horse

free logic: a logic which holds that the quantifiers of first-order logic are ontologically committing but denies that singular terms are devices of ontological commitment

H

heavyweight Platonism: the view that abstract objects are fundamental features of reality

hermeneutic nominalism: the view that mathematicians themselves do not take their mathematical sentences to be ontologically committing to mathematical objects

I

indispensability of mathematics: the necessity of using mathematical terms in describing the physical world

instantiation: the relation between a thing and its essence whereby the essence is particularized; the thing is an instance of its essence; *e.g.*, Paul and John instantiate or are instances of *humanity*

intelligible realm/world: in ancient Platonism, the realm of the Ideas; a static, conceptual realm
intensional context: a non-extensional context (see *extensional context*)
intentionality: about-ness; object-directedness; being *of* something
internal questions: questions posed in the language of an adopted linguistic framework
ideal: non-physical, conceptual
idealization: a fictitious entity in a scientific theory which real objects approximate, *e.g.*, ideal gases or ideal fluids

L

lightweight Platonism: the view that abstract objects merely serve as the semantic referents of certain terms and as values of the variables of quantification
linguistic framework: an object language involving terms for certain kinds of objects to which one is not ontologically committed when speaking from a standpoint outside the framework
literally true: true; not metaphorically true
logically prior: prior in the order of explanation

M

mereology: area of metaphysics which studies parts and wholes
meta-language: language used to talk about a lower level language
meta-ontological anti-realism: the view that there are no objective answers to certain ontological questions
meta-ontology: a higher level philosophical discipline about how to settle questions of ontology
metaphysics: branch of philosophy exploring reality in its various aspects
metaphysical: pertaining to metaphysics; beyond the physical world; what is ultimately real
metaphysical necessity/possibility: what must be actual/what could be actual
Middle Platonism: a school of philosophical thought which evolved from classical Platonism and flourished from the first

century B.C. until the third century. Middle Platonists identified Plato's Ideas with God's thoughts.

modal operator: a logical prefix which serves to give a modal status to a statement, like "Necessarily,..." or "Possibly,..."

N

nature: the essence of a thing, or those properties without which the thing would not exist

necessary: having to be certain way; being impossible to be otherwise

neo-Meinongianism: the view deriving from Alexius Meinong that there are (in a neutral sense) non-existent objects bearing properties

neo-Quinean: deriving from the view of the influential American philosopher W. V. O. Quine

neutralism: the view that existential quantification and use of singular terms are not ontologically committing

nominalism: in medieval debates, the view that universals do not exist; in the contemporary debate, the view either that abstract objects do not exist or that objects of a certain sort normally taken to be abstract, such as mathematical objects, do not exist

nominalistic content: the non-Platonistic contents of scientific theories

nominalistic scientific realism: the view that the nominalistic content of scientific theories is mostly true, while its Platonistic content is untrue

nominalization: finding a noun or noun phrase to express a notion; gerunds are often used for this purpose

O

ontological: pertaining to being or existence

ontological assay: an account of the metaphysical constituents of a thing

ontological commitment: something that one must regard as real or existent

ontological dependence: dependence in being (regardless of temporal beginning)

ontological dispute: a disagreement over what exists or is real

ontologically inflationary: carrying gratuitous ontological commitments

ontology: the study of what exists; an account of what exists

ontology, constituent: an ontology which takes things to be metaphysically composed of other entities; *e.g.*, a substratum and its properties

ontology, relational: an ontology which takes things to be metaphysically simple, *i.e.*, lacking any metaphysical constituents

P

paraphrastic strategies: anti-realist strategies for paraphrasing sentences involving quantification over or singular terms referring to abstract objects, so that their ontological commitments are eliminated

particular quantifier: an ontologically neutral existential quantifier

patristic: pertaining to the Church Fathers

Platonism: the view that there exist abstract objects

possible worlds: ways reality might have been; maximal states of affairs

postulationalism: the view which treats the axioms of competing mathematical theories as postulates whose consequences may be explored

presentism: the view that the present is ontologically privileged; the past and future do not exist

pretense theory: a theory of fiction according to which statements of fiction are prescribed to be imagined true

property: a universal quality which is exemplified by particulars

property instance: the particular exemplification of a property, as the redness of this rose; property instances are particulars, not universals

proposition: the information content expressed by sentences

Q

quantification: *see* existentially quantified statement, universally quantified statement

quantifier: *see* existential quantifier, universal quantifier

quantify over (something): to take (something) as the value of the variable lying within the scope of the quantifier

R

realism: the view with respect to a certain thing that that thing exists

referent: what one refers to

referentialism: doctrine that singular terms successfully refer only if there are mind-independent objects in the world which are the denotations or referents of the terms

reify: to construe something as thing by nominalizing an expression for it; *e.g.*, if one hunts deer, one might say that one is engaged in deer-hunting

revolutionary nominalism: the view that mathematics needs to be revised in such a way that its sentences are not ontologically committing to abstract objects

S

self-existent: existing independently of anything else

semantic ascent: talking *about* a claim rather than simply *making* the claim itself

semantic descent: simply *making* a claim rather than talking *about* the claim

sensible realm/world: in ancient Platonism, the realm of temporal becoming and concrete objects

simple: having no parts; uncomposed

simplicity: the property of being simple

singular term: a word (or words) which serves to pick out a specific individual, typically proper names like "John" or "U.S.S. Constitution," definite descriptions like "the coin in my pocket" or "your left arm," and demonstratives like "this" and "that"

sole ultimate reality: the only self-existent, uncreated thing

substances: things which exist in their own right and not as modifications of something else

T

temporal: located in time

tense operator: a logical prefix which serves to give a tenseless statement a tense, like "It was the case that..." or "It will be the case that..."

tensed theory of time: a view of time which holds that the tense of moments or events in time is objective and so the distinction between past, present, and future is an objective feature of the world

tenseless: lacking any relation to the present and so neither past, present, nor future

tenseless theory of time: a view of time which implies that all moments or events in time are equally real and the distinction between past, present, and future is merely a subjective feature of consciousness

theoretical entity: an unobservable object which may or may not exist but is posited by a scientific theory for its value in modeling physical phenomena

transcendent: beyond space and time, or beyond the physical universe

U

universal (*n.*): a property; it is what particular objects have in common that explains their resemblance to one another; *e.g.*, a British pillar box and a fire engine both share the universal *redness*

Universal Instantiation: a logical rule permitting us to infer from "Everything is *F*" that "*a* is *F*"

universally quantified statement: an assertion about all the members of a class, usually indicated by words like "all," "every," "any," and so forth

universal quantifier: a logical operator which makes a statement a universally quantified statement. Ordinary language expressions like "all," "every," "any," and so forth, are informal quantifiers; in formal or symbolic logic the operator "$\forall$" is a formal quantifier symbolizing such informal expressions. The symbol for the formal quantifier is often omitted for simplicity's sake. For example, instead of $(\forall x)\ (Fx \rightarrow Gx)$, one might write $(x)\ (Fx \rightarrow Gx)$, to be read "For any x, if x is F, then x is G."

univocal: having a single meaning; not equivocal

V

value: in logic the thing that replaces the variable in a sentence like $(\exists x)$ (x is an F); *e.g.*, in "*a* is *F*" the thing *a* stands for is the value of x.

variable: in first-order logic, the universal and existential quantifiers "∀" and "∃" are said to bind or have within their scope variables (like *x*, *y*, *z*) which may be replaced by certain constants (like *a*, *b*, *c*) to make statements about certain individuals in the domain of objects. *E.g.*, in $(\forall x)\ (Fx \rightarrow Gx)$, *x* is a variable which maybe replaced by the constant *a* to make the statement $Fa \rightarrow Ga$.

Z

Zermelo-Fraenkel set theory: standard set theory whose axioms were laid by Ernst Zermelo and Abraham Fraenkel; ZFC is Zermelo-Fraenkel set theory plus the Axiom of Choice

Glossary

Peter van Inwagen

Abstract Object: In this phrase, the word 'object' is used in a very general sense, so general that *everything* is an object. Any definition of 'abstract object' would be controversial, but it is generally agreed that numbers (and other mathematical objects) and propositions and properties and relations are abstract objects. Anything that is not an abstract object is a *concrete* object.

Antecedent (of a pronoun): A word or phrase in a sentence that a pronoun "refers back to." In the following two sentences, the bold-face phrase is the antecedent of the italicized pronoun.

> The dog barked at **the cat** and chased *it*.
> My dog barks at **cats** and chases *them*.

Antecedent (of a conditional statement): See *Conditional statement.*

Argument: A series of statements designed to show that a certain proposition (the *conclusion* of the argument) follows logically from certain other propositions (the *premises* of the argument). If the conclusion of an argument does follow logically from its premises, the argument is said to be *logically valid*—or simply *valid*. If the conclusion of an argument follows logically from its premises *and* those premises are all true, the argument is said to be *sound*.

Aristotelianism: In its most general sense, this word refers to the philosophy of Aristotle. In discussions of the problem of universals, it refers to the doctrine that universals are universals *in rebus*.

Aseity: To say that an entity has aseity is to say that it exists "in itself" (Latin: *a se*)—that is, that its existence does not depend, in any degree, however small, on any other being.

Conclusion: See *Argument.*

Concrete Object: See *Abstract object.*

Conditional Statement: An "if-then" statement. Conditional statements are of two types, illustrated by 'If Booth didn't shoot Lincoln, someone else did' and 'If Booth hadn't shot Lincoln, someone else would have'. Conditionals of the latter sort have been called counterfactual conditionals, *subjunctive* conditionals (Craig), and *'were'-'would'* conditionals (van Inwagen). The conditional 'If Booth hadn't shot Lincoln, someone else would have' is equivalent to 'If it were the case that Booth didn't shoot Lincoln, it would be the case that someone else shot Lincoln'. In virtue of this equivalence, 'Booth didn't shoot Lincoln' is called the *antecedent* of 'If Booth hadn't shot Lincoln, someone else would have' and 'Someone else would have' is called its *consequent.*

Consequent: See *Conditional statement.*

Count-noun: In English, a noun that has a plural form and can be modified by an indefinite article ('a'/'an'). For example, 'cat' is a count-noun because one can speak of 'cats' and of 'a cat'. (Almost all nouns *can* be used as count nouns. 'Butter' is not a count-noun if it denotes a certain stuff, and 'democracy' is not a count-noun if it denotes a certain form of government—but 'butters' means '*kinds* of butter' and 'a democracy' means 'a *country* whose form of government is democracy').

Definiendum: See *Definition.*

Definiens: See *Definition.*

Definition: A statement of the meaning of a word or phrase. If the statement is 'Triangle' means 'polygon with three edges and three vertices' the whole statement is the *definition* and 'triangle' is the *definiendum* of the definition and 'polygon with three edges and three vertices' is its *definiens.*

Descriptive Identity: See *Identity.*

Enthymeme: An argument that is not technically logically valid, but which would be if one or more missing premises were supplied—premises regarded by the person offering the argument as "too obvious to be worth stating" and therefore called *suppressed* premises.

Existential Generalization: A rule of logical inference. The following two inferences are examples of existential generalization: Socrates is mortal; *therefore* Someone is mortal; The shadow of the flagpole moved across the lawn; *therefore* Something moved across the lawn,

Existential Quantifier: See *Quantifiers, the.*

Explanatory Principle, The: A principle that is the generalization of these examples: the apple is green because (or in virtue of the fact that) it has the property greenness; the coin is round because (or in virtue of the fact that) it has the property roundness.

"Expressed by": See *Proposition.*

Extension: See *Open sentence.*

Fathers of the Church: Christian theologians primarily of the fourth and fifth centuries who systematized Christian theology and defined Christian theological orthodoxy.

Free Variable: See *Quantifiers.*

Identity: *Numerical* identity is the relation that everything bears to itself and can bear to nothing else: for x to stand in the relation of numerical identity to y is for x and y to be one and the same thing. *Descriptive* identity the relation that everything bears to those things that are exactly like it and can bear to nothing else: for x to stand in the relation of descriptive identity to y is for x and y to be indistinguishable—or at least to be distinguishable only by location.

Lightweight Platonism: (1) Van Inwagen's name for his own position on the nature of abstract objects. It implies that abstract objects exist necessarily and that they lack causal powers. (2) Craig's name for the position of Hale, Dummett, and Wright on abstract objects.

Logical validity: If the conclusion of an argument can be deduced by correct reasoning from its premises, the argument is logically valid. An argument can be logically valid if some of or all its premises are false.

Meinonginanism: The philosophy of the Austrian philosopher Alexius Meinong (1853–1920), and especially its central tenet: that, since every thought has an object, and since there is no such thing as (for example) the fountain of youth, some thoughts (such as the thought "The fountain of youth is a fable") have as their objects things such that there are no such things. *Neo*-Meinonginanism is the thesis that there are things that do not exist.

Meta-ontology: A field of study that enquires into the nature and scope of ontology.

Neo-Meinonginanism: See *Meinonginanism.*

Nicene Creed, the: A creed that is recited in Eastern Orthodox, Roman Catholic, and Anglican Eucharistic liturgies. Its name derives from the fact that it has been traditionally associated with the Council of Nicea or Nicaea (modern İznik in Turkey). The Council was held in A.D. 325.

Nominalism: In this volume, the doctrine that there are no *abstract objects.*

Numbers: For the purposes of this volume, abstract objects that serve as the measures of things; for example the number 67 is the size of the set whose members are the counties of Florida, and the (positive) square root of 2 measures the ratio of the length of the diagonal of a square to the length of one of its sides.

Numerical identity: See *Identity.*

Objectual quantification: See *Quantification.*

Ontological commitment: A person's ontological commitments comprise those objects whose existence is a logical consequence of the totality of his or her beliefs.

Ontology: (1) A part of philosophy; the study that enquires into the nature of being and non-being and existence and non-existence. (2) A part of philosophy; the study that attempts to answer the question, "What is there?" (2) Any given answer to the question, "What is there?"

Ontology Room, the: A trope used by van Inwagen. To say that a conversation takes place "in the Ontology Room" is to say that all parties to that conversation have agreed to certain conventions concerning "ontological" words and phrases (like 'exist', 'be', 'there is'), the most important being that these words be expressed in terms of quantifiers and variables or (what is the same thing) in Revised English.

Open sentence: A sentence in which there is at least one occurrence of a free variable (e.g., '*x* is wise'). We are concerned only with the case in which one variable occurs free. (In the sentence '*x*'s mother is wiser than *x*' there are two *occurrences* of a free variable, but only one variable—namely, '*x*'—is free in that sentence.) Anything that is wise is said to *satisfy* the open sentence '*x* is wise', and anything less wise than its mother satisfies the open sentence '*x*'s mother is wiser than *x*'. The set of things that satisfy an open sentence is called its *extension.*

Particular Quantifier: Most philosophers agree that it would be wrong to symbolize 'Some of the former employers listed in his resumé do not exist' as '$\exists x$ (*x* is listed in his resumé as a former employer, and *x* does not exist)'. But if not that way, how? One suggestion is to employ substitutional quantification: 'Σx

(x is listed in his resumé as a former employer, and x does not exist)'—which follows from 'The Nantasket Brass Basket and Glass Casket Corporation is listed in his resumé as a former employer, and the Nantasket Brass Basket and Glass Casket Corporation does not exist'.) Anyone who accepts this suggestion will not be willing to call 'Σ' "the existential quantifier" (or even "the substitutional existential quantifier"). The name generally given to 'Σ' is "the particular quantifier."

Platonism: In its most general sense, this word refers to the philosophy of Plato. In discussions of the problem of universals, it refers to the doctrine that universals are universals *ante res*. Platonism is also called Realism and Platonic Realism. See also *Lightweight Platonism*.

Premise: See *Argument*.

Property: An attribute, quality, feature, character, or trait.

Proposition: A thing that can be true or false; a thing that can be believed and can be *stated* by the use of a declarative sentence. If someone says, "Albany is the capital of the state of New York," that person *states that* Albany is the capital of New York, and no doubt *believes that* Albany is the capital of New York. Moreover, it is *true* that Albany is the capital of New York. The proposition that Albany is the capital of New York is the thing that (in this example) is stated, is believed, and is true.

Quantification: The aspect of sentence structure that marks out sentences that express generalizations of these two sorts: (1) Universal generalizations like 'All Greeks are mortal' and 'Every Greek is mortal', and (2) Particular generalizations like 'Some mortals are Greeks' and 'There are Greek mortals'. The term 'quantification' is sometimes understood in a way that allows two kinds of quantification, *objectual* quantification, and *substitutional* quantification. If '$\exists$' is the objectual "existential" quantifier, then, e.g., '$\exists x$ x is a dragon' is true just in the case that some object satisfies the open sentence 'x is a dragon' (see *Open sentence*). If 'Σ' is its substitutional counterpart, the so-called particular quantifier, then 'Σx x is a dragon' is true just in the case that some name may be substituted for 'x' in the open sentence 'x is a dragon' so as to produce a truth. Thus, if the sentence 'Fafnir is a dragon' is true, 'Σx x is a dragon' is true, and (because no object satisfies 'x is a dragon'), '$\exists x$ x is a dragon' is false.

Quantifiers: In the narrowest sense, two formal operators, the "universal" quantifier and the "existential" quantifier, often written '∀' and '∃', that turn open sentences into closed sentences. Thus, prefixing '∀*x*' to the open sentence 'if *x* is Greek, then *x* is mortal' produces a closed sentence that means 'All Greeks are mortal', and prefixing '∃*x*' to the open sentence '*x* is Greek and *x* is mortal' produces a closed sentence that means 'Some Greeks are mortal'. In van Inwagen's broader sense, the expressions 'everything is such that' and 'something is such that', the natural-language analogues of '∀' and '∃'. In the sentence '∃*x* (*x* is Greek and *x* is mortal)', '∃' is a quantifier, '∃*x*' is a quantifier *phrase*', and the round backets indicate the *scope* of that quantifier phrase. See also *Particular quantifier, Quantification, Specific quantifier, Variables.*

Realism: See *Platonism.*

Relation: If a proposition is what is expressed by a closed sentence (the sentence 'Montreal is a city' expresses the proposition that Montreal is a city), and if a property (or quality or attribute) is what is expressed by an open sentence in which one variable is free (the open sentence '*x* is a city' expresses the property of being an *x* such that *x* is a city), a relation is what is expressed by an open sentence in which two variables are free (the open sentence '*x* is to the north of *y*' expresses the relation "being to the north of."

Revised English: An artificially restricted and rigidified modification of English that van Inwagen uses to explain quantification. It is also the language of the Ontology Room.

Satisfaction: See *Open sentence.*

Scope: See *Quantifiers, the.*

Sound argument: See *Argument.*

Specific quantifier, the: Van Inwagen's term for the phrase 'something is such that'.

Substitutional quantification: See *Quantification.*

Universal: (noun) A thing that is universal to the members of a collection and defines membership in that collection. For example, the property greenness belongs to all members of, and only to the members of, class of green things, and the novel *War and Peace* is common to all its tangible copies.

Universal *ante res*: A universal as Platonists understand universals: a universal that exists independently of the objects it characterizes, and would exist even if it characterized nothing. For example, if greenness is a universal *ante res*, it exists independently of green things and would exist even if nothing were green.

Universal *in rebus*: A universal as Aristotelians understand universals: a universal that is present in the objects it characterizes, and whose mode of existence is presence in those objects. If greenness is a universal *in rebus*, it can exist only as a presence in green things.

Universal instantiation: A logical rule of inference illustrated by the inferences 'All Greeks are mortal. *Therefore*, If Socrates is Greek, Socrates is mortal' and 'God has created everything besides himself. *Therefore*, If Socrates is not God, God has created Socrates'.

Universal quantifier: See *Quantifiers*.

Variable: Variables are the italicized letters from the end of the roman alphabet in expressions like '$\exists x$ (x is Greek and x is mortal)' and '$\forall z \exists y$ (z loves y)'. Variables are essentially third-person-singular pronouns: '$\exists x$ (x is Greek and x is mortal)' is an abbreviation of 'Something^{x} is such that (it^{x} is Greek and it^{x} is moral)' and '$\forall z \exists y$ (z loves y)' is an abbreviation of 'Everything^{z} is such that (something^{y} is such that (it^{z} loves it^{y}))'.

'Were'-'would' Conditional: See *Conditional statement*.

Bibliography

Alston, William. "Audi on Nondoxastic Faith." In *Rationality and the Good: Critical Essays on the Ethics and Epistemology of Robert Audi*, edited by Mark Timmons, John Greco, and Alfred R. Mele, 123–40. Oxford: Oxford University Press, 2007.

Alston, William. "Belief, Acceptance, and Religious Faith." In *Faith, Freedom, and Rationality*, edited by Jeff Jordan and Daniel Howard-Snyder, 3–27. Lanham, MD: Rowman & Littlefield, 1996.

Ayer, Alfred J. *Thinking and Meaning*. London: H. K. Lewis, 1947.

Azzouni, Jody. *Deflating Existential Consequence: A Case for Nominalism*. Oxford: Oxford University Press, 2004.

Azzouni, Jody. "Evading Truth Commitments: The Problem Reanalyzed." *Logique et Analyse* 52, no. 206 (2009): 139–76.

Azzouni, Jody. "On 'On What There Is'." *Pacific Philosophical Quarterly* 79 (1998): 1–18.

Azzouni, Jody. "Ontological Commitment in the Vernacular." *Noûs* 41, no. 2 (2007): 204–26.

Baker, Alan. "Are There Genuine Mathematical Explanations of Physical Phenomena?" *Mind* 114, no. 454 (2005): 223–38.

Baker, Lynne Rudder. *The Metaphysics of Everyday Life: An Essay in Practical Realism*. Cambridge: Cambridge University Press, 2007.

Balaguer, Mark. *Platonism and Anti-Platonism in Mathematics*. New York: Oxford University Press, 1998.

Balaguer, Mark. "Platonism in Metaphysics." In *The Stanford Encyclopedia of Philosophy*. Summer 2009 ed. Article published June 21, 2009. http://plato.stanford.edu/archives/sum2009/entries/platonism/.

Båve, Arvid. "A Deflationary Theory of Reference." *Synthèse* 169, no. 1 (2009): 51–73.

Brentano, Franz. "The Distinction between Mental and Physical Phenomena." In *Realism and the Background of Phenomenology*, edited by Roderick M. Chisholm. Translated by D. B. Terrell, 39–61. Atascadero, CA: Ridgeview, 1960a.

Brentano, Franz. "Genuine and Fictitious Objects." In *Realism and the Background of Phenomenology*, edited by Roderick M. Chisholm. Translated by D. B. Terrell, 71–75. Atascadero, CA: Ridgeview, 1960b.

Burgess, John P. "Mathematics and *Bleak House*." *Philosophia Mathematica* 12, no. 1 (2004): 18–36.

Burgess, John P. "Why I Am Not a Nominalist." *Notre Dame Journal of Formal Logic* 24, no. 1 (1983): 93–105.

Burgess, John P., and Gideon A. Rosen. *A Subject with No Object: Strategies for Nominalistic Interpretation of Mathematics*. Oxford: Oxford University Press, 1997.

Carnap, Rudolf. *Meaning and Necessity: A Study in Semantics and Modal Logic*. Chicago, IL: University of Chicago Press, 1956.

Chalmers, David J. "Ontological Anti-realism." In *Metametaphysics: New Essays on the Foundations of Ontology*, edited by David J. Chalmers, David Manley, and Ryan Wasserman, 77–129. Oxford: Oxford University Press, 2009.

Chihara, Charles S. "Nominalism." In *The Oxford Handbook of Philosophy of Mathematics and Logic*, edited by Stewart Shapiro, 483–514. Oxford: Oxford University Press, 2005.

Chihara, Charles S. *Ontology and the Vicious Circle Principle*. Ithaca, NY: Cornell University Press, 1973.

Chihara, Charles S. *A Structural Account of Mathematics*. Oxford: Clarendon Press, 2004.

Chihara, Charles S. *The Worlds of Possibility: Modal Realism and the Semantics of Modal Logic*. Oxford: Clarendon Press, 1998.

Craig, William Lane. *God and Abstract Objects: The Coherence of Theism: Aseity*. Cham: Springer, 2017.

Craig, William Lane. *God over All: Divine Aseity and the Challenge of Platonism*. Oxford: Oxford University Press, 2016.

Craig, William Lane. *The Tensed Theory of Time: A Critical Examination*. Synthèse Library 293. Dordrecht: Kluwer Academic, 2000a.

Craig, William Lane. *The Tenseless Theory of Time: A Critical Examination*. Synthèse Library 294. Dordrecht: Kluwer Academic, 2000b.

David K. Lewis and Stephanie Lewis, "Holes," *Australasian Journal of Philosophy* 48/2 (1970): 206–212.

Dummett, Michael. *Frege: Philosophy of Mathematics*. Cambridge, MA: Harvard University Press, 1991.

Dummett, Michael. "Platonism." In *Truth and Other Enigmas*, by Michael Dummett, 202–14. Cambridge: Harvard University Press, 1978.

Einstein, Albert. "Letter to Maurice Solovine, March 30, 1952". In *Letters to Solovine*, edited by Albert Einstein. Translated by Wade Baskin, 132–33. New York: Philosophical Library, 1987.

Eklund, Matti. "Carnap and Ontological Pluralism." In *Metametaphysics: New Essays on the Foundations of Ontology*, edited by David J.

Chalmers, David Manley, and Ryan Wasserman, 130–56. Oxford: Oxford University Press, 2009.

Eklund, Matti. "The Picture of Reality as an Amorphous Lump." In *Contemporary Debates in Metaphysics*, edited by Theodore Sider, John Hawthorne, and Dean W. Zimmerman, 382–96. Oxford: Blackwell, 2008.

Field, Hartry H. *Science without Numbers: A Defence of Nominalism*. Princeton, NJ: Princeton University Press, 1980.

Findlay, J. N. *Meinong's Theory of Objects and Values*. 2nd ed. Oxford: Clarendon Press, 1963.

Franklin, James. *An Aristotelian Realist Philosophy of Mathematics: Mathematics as the Science of Quantity and Structure*. Basingstoke, Hampshire: Palgrave Macmillan, 2014.

Frege, Gottlob. *The Foundations of Arithmetic: A Logico-Mathematical Enquiry into the Concept of Number*, translated by J. L. Austin. 2nd rev. ed. Evanston, IL: Northwestern University Press, 1968.

Grattan-Guinness, Ivor. "Solving Wigner's Mystery: The Reasonable (Though Perhaps Limited) Effectiveness of Mathematics in the Natural Sciences." *Mathematical Intelligencer* 30, no. 3 (2008): 7–17.

Hellman, Geoffrey. *Mathematics without Numbers: Towards a Modal-Structural Interpretation*. Oxford: Clarendon Press, 1989.

Hellman, Geoffrey. "On Nominalism." *Philosophy and Phenomenological Research* 62, no. 3 (2001): 691–705.

"Holes" in David Lewis, *Collected Philosophical Papers, Vol.* I, 3-9. New York and Oxford: Oxford University Press, 1983.

Hofweber, Thomas. "Ontology and Objectivity." Ph.D. Dissertation, Stanford, CA: Stanford University, 1999.

Hudson, Hud. "Confining Composition." *The Journal of Philosophy* 103, no. 12 (2006): 631–51.

Husserl, Edmund. *Logical Investigations*. Translated by J. M. Findlay. 2 vols. New York: Humanities Press, 1970.

Inman, Ross. "On Christian Theism and Unrestricted Composition." Paper presented at the Annual Meeting of the Evangelical Philosophical Association, Milwaukee, WI, November, 2012.

Kripke, Saul A. "Speaker's Reference and Semantic Reference." In *Contemporary Perspectives in the Philosophy of Language*, edited by Peter A. French, Theodore E. Uehling, Jr. and Howard K. Wettstein, 6–27. Minneapolis: University of Minnesota Press, 1979.

Lambert, Karel. "Existential Import Revisited." *Notre Dame Journal of Formal Logic* 4, no. 4 (1963): 288–92.

Lambert, Karel. *Free Logic: Selected Essays*. Cambridge: Cambridge University Press, 2003.

Lambert, Karel. *Meinong and the Principle of Independence*. Cambridge: Cambridge University Press, 1983.

Lambert, Karel. "The Nature of Free Logic." In *Philosophical Applications of Free Logic*, 3–14. Oxford: Oxford University Press, 1991.

Lange, Marc. "What Makes a Scientific Explanation Distinctively Mathematical?" *British Journal for the Philosophy of Science* 64, no. 3 (2013): 485–511.

Leftow, Brian. *God and Necessity*. Oxford: Oxford University Press, 2012.

Leng, Mary. *Mathematics and Reality*. Oxford: Oxford University Press, 2010.

Lewis, David K. *Parts of Classes*. Oxford: Basil Blackwell, 1991.

Lewis, David K. "Tensed Quantifiers." In *Oxford Studies in Metaphysics*, edited by Dean Zimmerman, 1: 3–14. Oxford: Oxford University Press, 2004.

Lewis, David and Stephanie Lewis. "Holes." In *Collected Philosophical Papers*, edited by David Lewis, vol. I, 3–9. New York and Oxford: Oxford University Press, 1983.

Maddy, Penelope. "Believing the Axioms. I." *The Journal of Symbolic Logic* 53, no. 2 (1988): 481–511.

Maddy, Penelope. "Believing the Axioms. II." *The Journal of Symbolic Logic* 53, no. 3 (1988): 736–64.

Maddy, Penelope. *Defending the Axioms: On the Philosophical Foundations of Set Theory*. Oxford: Oxford University Press, 2011.

Maddy, Penelope. *Naturalism in Mathematics*. Oxford: Clarendon Press, 1997.

Mag Uidhir, Christy. "Introduction: Art, Metaphysics, and the Paradox of Standards." In *Art and Abstract Objects*, edited by Christy Mag Uidhir, 1–26. Oxford: Oxford University Press, 2012.

Margolis, Joseph. "Reference as Relational: *Pro* and *Contra*." In *Nonexistence and Predication*, edited by Rudolf Haller. Grazer Philosophische Studien, 327–58. Amsterdam: Rodopi, 1986.

Maudlin, Tim. "On the Foundations of Physics." July 5, 2013, 3:16, https://www.3-16am.co.uk/articles/on-the-foundations-of-physics.

Menzel, Christopher. "God and Mathematical Objects." In *Mathematics in a Postmodern Age*, edited by Russell W. Howell and W. James Bradley, 65–97. Grand Rapids, MI: Eerdmans, 2001.

Menzel, Christopher. "Theism, Platonism, and the Metaphysics of Mathematics." *Faith and Philosophy* 4, no. 4 (1987): 365–82.

Merricks, Trenton. *Truth and Ontology*. Oxford: Clarendon Press, 2007.

Morris, Thomas V., and Christopher Menzel. "Absolute Creation." *American Philosophical Quarterly* 23, no. 4 (1986): 353–62.

Naylor, Margery Bedford. "A Note on David Lewis's Realism about Possible Worlds." *Analysis* 46, no. 1 (1986): 28–29.

Nolt, John. "Free Logics." In *Philosophy of Logic*, edited by Dale Jacquette. Handbook of the Philosophy of Science, 5: 1023–1060. Amsterdam: Elsevier, 2006.

Oppy, Graham. "Response to Greg Welty." In *Beyond the Control of God?: Six Views on the Problem of God and Abstract Objects*, edited by Paul M. Gould. Bloomsbury Studies in Philosophy of Religion, 104–6. London: Bloomsbury, 2014.

Orenstein, Alex. "Is Existence What Existential Quantification Expresses?" In *Perspectives on Quine*, edited by Robert B. Barrett and Roger F. Gibson, 245–70. Oxford: Basil Blackwell, 1990.

Parsons, Terence. *Nonexistent Objects*. New Haven, CT: Yale University Press, 1980.

Perszyk, Kenneth J. *Nonexistent Objects: Meinong and Contemporary Philosophy*. Nijhoff International Philosophy Series 49. Dordrecht: Kluwer Academic Publishers, 1993.

Philo of Alexandria. *On the Creation of the Cosmos according to Moses*. Translated, with an Introduction and Commentary by David T. Runia. Philo of Alexandria Commentary Series 1. Leiden: Brill, 2001.

Plantinga, Alvin. "Response to William Lane Craig's Review of *Where the Conflict Really Lies*." *Philosophia Christi* 15, no. 1 (2013): 175–82.

Plantinga, Alvin. "Theism and Mathematics." *Theology and Science* 9, no. 1 (2011): 27–33.

Plantinga, Alvin. *Where the Conflict Really Lies: Science, Religion, and Naturalism*. Oxford: Oxford University Press, 2011.

Plato. *Timaeus and Critias*. Translated with an Introduction and Appendix on *Atlantis* by H. D. P. Lee, Penguin Classics. Harmondsworth, Middlesex, England: Penguin Books, 1965.

Potter, Michael. *Set Theory and Its Philosophy: A Critical Introduction*. Oxford: Oxford University Press, 2004.

Priest, Graham. "The Closing of the Mind: How the Particular Quantifier Became Existentially Loaded Behind Our Backs." *The Review of Symbolic Logic* 1, no. 1 (2008): 42–55.

Priest, Graham. *Towards Non-being: The Logic and Metaphysics of Intentionality*. Oxford: Oxford University Press, 2005.

Putnam, Hilary. "What Is Mathematical Truth?" In *Philosophical Papers*. Vol. 1, *Mathematics, Matter, and Method*, edited by Hilary Putnam, 2nd ed., 60–78. Cambridge: Cambridge University Press, 1979.

Quine, W. V. O. "On What There Is." *The Review of Metaphysics* 2, no. 5 (1948): 21–38.

Quine, W. V. O. *Philosophy of Logic*. 2nd ed. Cambridge, MA: Harvard University Press, 1986.

Rosen, Gideon, and John P. Burgess. "Nominalism Reconsidered." In *The Oxford Handbook of Philosophy of Mathematics and Logic*, edited by Stewart Shapiro, 515–35. Oxford: Oxford University Press, 2005.

Routley, Richard. *Exploring Meinong's Jungle and Beyond: An Investigation of Noneism and the Theory of Items*. Canberra: Research School of Social Sciences, Australian National University, 1979.

Schroeder, Manfred R. "The Unreasonable Effectiveness of Number Theory in Physics, Communication and Music." *Proceedings of Symposia in Applied Mathematics* 46 (1992): 1–19.

Searle, John R. *The Construction of Social Reality*. New York: Free Press, 1995.

Searle, John R. *Expression and Meaning: Studies in the Theory of Speech Acts*. Cambridge: Cambridge University Press, 1979.

Searle, John R. *Intentionality: An Essay in the Philosophy of Mind*. Cambridge: Cambridge University Press, 1983.

Searle, John R. *Speech Acts: An Essay in the Philosophy of Language*. Cambridge: Cambridge University Press, 1969.

Sider, Theodore. "Presentism and Ontological Commitment." *The Journal of Philosophy* 96, no. 7 (1999): 325–47.

Sider, Theodore. "Quantifiers and Temporal Ontology." *Mind* 115, no. 457 (2006): 75–97.

Sider, Theodore. *Writing the Book of the World*. Oxford: Clarendon Press, 2011.

Smith, R. Scott. *Naturalism and Our Knowledge of Reality: Testing Religious Truth-Claims*. Farnham, England: Ashgate, 2012.

Van Inwagen, Peter. "Being, Existence, and Ontological Commitment." In *Metametaphysics: New Essays on the Foundations of Ontology*, edited by David J. Chalmers, David Manley, and Ryan Wasserman, 472–506. Oxford: Clarendon Press, 2009.

Van Inwagen, Peter. "Did God Create Shapes?" *Philosophia Christi* 17, no. 2 (2015): 285–90.

Van Inwagen, Peter. "The Doctrine of Arbitrary Undetached Parts." *Pacific Philosophical Quarterly* 62 (1981): 123–37.

Van Inwagen, Peter. *Existence: Essays in Ontology*. Cambridge: Cambridge University Press, 2014.

Van Inwagen, Peter. "Fiction and Metaphysics." *Philosophy and Literature* 7, no. 1 (1983): 67–77.

Van Inwagen, Peter. "God and Other Uncreated Things." In *Metaphysics and God: Essays in Honor of Eleonore Stump*, edited by Kevin Timpe, 3–20. London: Routledge, 2009.

Van Inwagen, Peter. *Material Beings*. Ithaca, NY: Cornell University Press, 1990.

Van Inwagen, Peter. "Meta-ontology." In *Ontology, Identity, and Modality: Essays in Metaphysics*. Cambridge Studies in Philosophy, 13–31. Cambridge: Cambridge University Press, 2001.

Van Inwagen, Peter. *Ontology, Identity, and Modality: Essays in Metaphysics*. Cambridge Studies in Philosophy. Cambridge: Cambridge University Press, 2001.

Van Inwagen, Peter. "Quantification and Fictional Discourse." In *Empty Names, Fiction, and the Puzzles of Non-existence*, edited by Anthony

Everett and Thomas Hofweber, 235–46. Stanford: Center for the Study of Language and Information, 2000.
Van Inwagen, Peter. "Response to William Lane Craig's *God over All.*" *Philosophia Christi* 21 (2019): 267–75.
Van Inwagen, Peter. "A Theory of Properties." In *Oxford Studies in Metaphysics*, edited by Dean Zimmerman, 1: 107–38. Oxford: Oxford University Press, 2004.
Varzi, Achille C. "Words and Objects." In *Individuals, Essence and Identity: Themes of Analytic Metaphysics*, edited by Andrea Bottani, Massimiliano Carrara, and Pierdaniele Giaretta. Topoi Library 4, 49–75. Dordrecht: Kluwer Academic Publishers, 2002.
Vision, Gerald. "Reference and the Ghost of Parmenides." In *Non-existence and Predication*, edited by Rudolf Haller. Grazer Philosophische Studien, 297–326. Amsterdam: Rodopi, 1986.
Walton, Kendall L. *Mimesis as Make-Believe: On the Foundations of the Representational Arts*. Cambridge, MA: Harvard University Press, 1990.
Welty, Greg. "Theistic Conceptual Realism." In *Beyond the Control of God?: Six Views on the Problem of God and Abstract Objects*, edited by Paul M. Gould, 81–96. London: Bloomsbury, 2014.
Welty, Greg. "Theistic Conceptual Realism: The Case for Interpreting Abstract Objects as Divine Ideas." D.Phil. thesis, University of Oxford, 2006.
Wigner, Eugene P. "The Unreasonable Effectiveness of Mathematics in the Natural Sciences." *Communications on Pure and Applied Mathematics* 13, no. 1 (1960): 1–14.
Willard, Dallas. "For Lack of Intentionality." In *Phenomenology 2005, Selected Essays from North America* (Part 2), edited by Lester Embree and Thomas Nenon, 5: 593–612. Bucharest: Zeta Books, 2007.
Willard, Dallas. *Logic and the Objectivity of Knowledge*. Athens: Ohio University Press, 1984.
Williamson, Timothy. *Modal Logic as Metaphysics*. Oxford: Oxford University Press, 2013.
Yablo, Stephen. "Abstract Objects: A Case Study." In *Realism and Relativism*, edited by Ernest Sosa and Enrique Villanueva. Philosophical Issues, 220–40. Boston, MA: Blackwell, 2002.
Yablo, Stephen. "Does Ontology Rest on a Mistake?" *Proceedings of the Aristotelian Society, Supplementary Volume* 72 (1998): 229–61.
Yablo, Stephen. "Go Figure: A Path Through Fictionalism." In *Figurative Language*, edited by Peter A. French and Howard K. Wettstein. Midwest Studies in Philosophy, 25: 72–102. Oxford: Blackwell, 2001.
Yablo, Stephen. "The Myth of the Seven." In *Fictionalism in Metaphysics*, edited by Mark Eli Kalderon, 88–115. Oxford: Clarendon Press, 2005.
Yablo, Stephen. "A Paradox of Existence." In *Empty Names, Fiction, and the Puzzles of Non-Existence*, edited by Anthony Everett and Thomas

Hofweber, 275–312. Stanford, CA: Center for the Study of Language and Information, 2000.
Yagisawa, Takashi. "Beyond Possible Worlds." *Philosophical Studies* 53, no. 2 (1988): 175–204.
Yagisawa, Takashi. "Beyond Possible Worlds," *Philosophical Studies* 53 (1988): 175–204.

Bibliography

Adams, Robert M. "The Metaphysical Lightness of Being." Paper presented at the Carolina Metaphysics Workshop, Duck, NC, 26 June, 2012.

Asimov, Isaac. "The Feeling of Power." *If: Worlds of Science-Fiction*, 1958. https://urbigenous.net/library/power.html.

Ayer, A. J.. *Thinking and Meaning*. London: H. K. Lewis, 1947.

Azzouni, Jody. *Deflating Existential Consequence: A Case for Nominalism*. Oxford: Oxford University Press, 2004.

Chisholm, Roderick M. *Person and Object*. LaSalle, IL: Open Court, 1976.

Church, Alonzo. "Ontological Commitment." *The Journal of Philosophy* 55 (1958): 1008–14.

Craig, William Lane. "God and Abstract Objects." *Philosophia Christi* 17 (2015a): 269–76.

Craig, William Lane. *God Over All: Divine Aseity and the Challenge of Platonism*. Oxford: Oxford University Press, 2016.

Craig, William Lane. "Response to Bridges and Van Inwagen." *Philosophia Christi* 17 (2015b): 291–97.

Craig, William Lane. "Response to Van Inwagen and Welty." *Philosophia Christi* 21 (2019): 277–86.

Dummett, Michael. "Nominalism." In *Truth and Other Enigmas*, 38–49. Cambridge, MA: Harvard University Press, 1978.

Falguera, José L., Concha Martínez-Vidal, and Gideon Rosen. "Abstract Objects." In *The Stanford Encyclopedia of Philosophy*, edited by Edward N. Zalta, Summer 2022 edition, 2021. https://plato.stanford.edu/archives/sum2022/entries/abstract-objects/.

Geach, Peter. *The Virtues: The Stanton Lectures 1973–74*. Cambridge: Cambridge University Press, 1977.

Hale, Bob. *Abstract Objects*. Oxford: Basil Blackwell, 1987.

Hale, Bob, and Crispin Wright. "Nominalism and the Contingency of Abstract Objects," *The Journal of Philosophy* 89 (1992): 111–35.

Hofweber, Thomas. *Ontology and the Ambitions of Metaphysics*. Oxford: Oxford University Press, 2016.

Meinong, Alexius. "Über Gegenstandstheorie." In *Gesamtausgabe*, edited by Rudolf Haller, Rudolf Kindinger, and Roderick M. Chisholm, 7 vols, 1969–73. Graz: Akademische Druck und Verlagsanstalt, 1904 [An English translation "The Theory of Objects" is included in Roderick M. Chisholm (ed.), *Realism and the Background of Phenomenology*.Atascadero CA: 1960.]

Merricks, Trenton. *Objects and Persons*. Oxford: Oxford University Press, 2001.

Mill, John Stuart. *A System of Logic, Ratiocinative and Inductive: Being a Connected View of the Principles of Evidence and the Methods of Scientific Investigation*. London and New York: Longmans, 1843/1961.

Orenstein, Alex. *Existence and the Particular Quantifier*. Philadelphia, PA: Temple University Press, 1978.

Quine, Willard Van Orman. *Word and Object*. Cambridge: MIT Press, 1960.

Van Inwagen, Peter. "Did God Create Shapes." *Philosophia Christi* 17 (2015a): 285–90.

Van Inwagen, Peter. "Fictionalist Nominalism and Applied Mathematics." *The Monist* 97 (2014a): 479–502.

Van Inwagen, Peter. "God and Other Uncreated Things." In *Metaphysics and God: Essays in Honor of Eleonore Stump*, edited by Kevin Timpe, 3–20. London and New York: Routledge, 2009.

Van Inwagen, Peter. "Inside and Outside the Ontology Room." In *Existence: Essays in Ontology*, edited by P. van Inwagen, 2–14. Cambridge: Cambridge University Press, 2014b.

Van Inwagen, Peter. "A Reply to Dr Craig." *Philosophia Christi* 17 (2015b): 299–305.

Van Inwagen, Peter. "Response to William Lane Craig's *God Over All*." *Philosophia Christi* 21 (2019): 267–75.

Van Inwagen, Peter. "A Theory of Properties." In *Oxford Studies in Metaphysics*, edited by Dean Zimmerman, vol. 1, 107–38. 2004. Oxford: Oxford University Press

Van Inwagen, Peter. "Why I Don't Understand Substitutional Quantification." *Philosophical Studies* 39 (1981): 281–85.

Index